The Frontiers of
Theory Development in Physics

A Methodological Study in its Dynamical Complexity

Giridhari L. Pandit [1]

Hans Günter Dosch [2]

[1]University of Delhi South Campus *and* Centre for Ecological Economics & Natural Resources, Isec, Bangalore - 560072. India; E-mail: gpandit9@gmail.com

[2]Institut für Theoretische Physik der Universität Heidelberg Philosophenweg 16, 69120 Heidelberg, Germany; E-mail: h.g.dosch@thphys.uni-heidelberg.de

Contents

Foreword

It is a great honour and pleasure for me to write some words on this book by my two esteemed colleagues G. L. Pandit and H. G. Dosch. As theorists in particle physics we are usually occupied with urgent day to day problems which should be solved in order that a student can get his Ph. D. or that we can present a talk at some conference. With this book, however, we are invited to take a step back and get a wider perspective on problem and theory development in the sciences, in particular in physics. How and why did the great "revolutions" in the physical sciences come about? How are the problems in physics, asked at various times, determined by the "reigning" theories of the time? What are the philosophical implications of progress in the sciences? Are we aiming at one great unified theory which one day will explain "everything" or will the quest for a "fundamental theory" be open ended? In the book by Pandit and Dosch the reader finds all these and many related questions posed with great clarity. The answers given by the authors are corroborated by important case studies. This is very much in the good tradition of physics that statements must be validated by experimental facts. The novel perspectives of the authors on theory development in science may clash with the cherished views of some philosophical schools. It is my hope that this book will lead both philosophers and physicists to critically examine their own positions with respect to the fundamental questions posed above. The authors achieve to show up the scientific frontiers in the relation of philosophy and the physical sciences from within these disciplines. I wish the authors that their book will lead to new questions and to further developments at the frontier of philosophy and physics, where - as in physics itself, as stressed repeatedly in the book - we certainly should aim at unification.

Otto Nachtmann

1

Preface

As we were engaged in a discussion concerning the dynamical aspects of physical theory during 2005-2006, we decided to work on the present research project. We began with the aim of producing a research article which could be of interest to the philosophers and scientists active at the frontiers of research. From the very outset, we recognized how challenging it might be to work on the context *in* science as a context of scientific reasoning in its own right, which is effective in physics although it has not received the attention it deserves within the methodology and philosophy of science. More specifically, in order to realize our plan, we decided to explore the core-context guided problem and theory development in physics with reference to Hermann Weyl's (Weyl 1918a, 1918b, 1919, 1929) principle of gauge invariance, which is a guiding principle in the Standard Model of Particle Physics. By the time our article began taking shape, the methodological aspects of the problem in which we were interested began dawning on us, revealing its fascinating complexity in all its details. This was quite challenging, since it demanded not only more and more space for an extended text but much greater attention commensurate with various aspects of problem and theory development in physics. It turned out, especially, that if we were to fulfill our task in a nearly adequate way, it would be necessary to widen the range of discussion on physical theory to more aspects of the development of the Standard Model of Particle Physics. Accordingly, we decided to work over an extended period of time in order to produce a more detailed account in the form of a book, hoping that it would turn out to be interesting, even challenging, for the methodologist (philosopher) interested in the development of science, on the one hand, and for the scientist interested in the methodology of inquiry, on the other.

The chapters 1-5 explore the physical and methodological aspects of theory development in science with particular reference to the Standard Model of elementary particle physics. The philosophical and physical background of these considerations is presented in chapter 6 and 7, respectively. We hope that these later chapters do not only offer our readers the necessary physical and philosophical background but are also a crucial input to the understanding of our arguments for core-contextual methodological structuralism.

Heidelberg, July 2012

G.L. Pandit   H.G. Dosch

Acknowledgements

The authors want to thank all the colleagues at the Institut fuer Theoretische Physik, Universität Heidelberg, in particular Dr. Eduard Thommes, for providing an excellent atmosphere of support for our study in problem and theory development in science. During the beginning of the project, the Alexander von Humboldt-Stiftung, Bonn, made the discussion meetings between the two authors from two continents and two scientific communities possible. The authors would like to thank the Alexander von Humboldt-Stiftung for this support. One of the authors (H G D) is grateful to the University of Delhi South Campus for the hospitality extended to him during his two stays in Delhi.

It is a pleasure to acknowledge detailed comments and suggestions which were received from colleagues and friends on the early drafts of this work. We are particularly thankful to Professor Otto Nachtmann. He read through the various drafts and provided encouragement throughout the whole period of our study. Nicholas Maxwell (2005, 2006), University College London, offered detailed comments on core-contextual methodological structuralism, as set forth in this study. During 2005 and 2008, one of the authors (G L P) was invited by the Fondation de la Maison des Sciences de L'home, Paris, facilitating his discussions with Prof. Bernard d'Espagnat, Universite de Paris XI), resulting in a dialogue on philosophy and methodology of physics between them. The autors are thankful to Bernard d'Espagnat and Richard T. Hull (Emeritus Professor, Deptartment of Philosophy, State University of New York at Buffalo) for their going through the early drafts of the present study and for receiving their comments.

To Dr. Volker A. Munz (Institut fuer Geschichte, Institut fuer Philosophie, Karl-Franzens-Universitaet Graz), General Secretary of Austrian Ludwig Wittgenstein Society (ALWS), the authors wish to express thanks for permission for incorporating Pandit (1989) in Chapter 6 (Sec. 6.2) of this book. Elaine Griffiths, Lecturer at the

Seminar für Übersetzen und Dolmetschen am Institut für Allgemeine und Angewandte Sprachwissenschaft der Universität Heidelberg very kindly took the trouble of checking some of the English by going through the whole manuscript. The authors would like to express their thanks to her. We also thank Mrs. Manuela Wirschke, Institut fuer Theoretische Physik, who was TeX-setting large parts of the book. Finally, the authors feel deeply indebted to their respective close relatives and friends across India and Germany for receiving abundant moral support during the course of their work on this project.

Chapter 1

METHODOLOGICAL STRUCTURALISM

Es ist häufig mit besonderer Betonung darauf hingewiesen worden, dass die Quantenmechanik es nur mit prinzipiell beobachtbaren Größen und nur mit physikalisch sinnvollen Fragen zu tun hat. Das ist gewisse zutreffend, es darf aber nicht speziell der Quantentheorie von vornherein als ein besonderer Vorzug gegenüber anderen Theorien angerechnet werden. Denn die Entscheidung darüber, ob eine physikalische Größe prinzipiell beobachtbar ist oder ob eine gewisse Frage einen physikalischen Sinn hat, lässt sich niemals a priori, sondern immer erst vom Standpunkt einer bestimmten Theorie aus treffen. Der Unterschied der verschiedenen Theorien liegt eben darin, dass nach der einen Theorie eine gewisse Größe prinzipiell beobachtbar, eine gewisse Frage physikalisch sinnvoll ist, nach der andern nicht. Max Planck (1965, 221-222)

Within science studies such as philosophy of science, physical theory has received much attention in those aspects which are not only celebrated but widely applied in different fields since the time of Newton. Given a physical theory T, in its field it must fulfil the two most important tasks of explanation and prediction of observable phenomena. As regards T's predictive power, what matters most is how successful T is empirically or experimentally, particularly with regard to the question how well-confirmed a theory T is in the light of experimental results. But as regards its explanatory power the relationship between theory and experiment gets much more complex, particularly when it is possible to consider it in relation to a rival theory Tr. Here it is the simplicity of its fundamental equations which plays a decisive role in theory-choice. Of course, using appropriate experimental facilities, it is with its mathematical structure that T is able to fulfil these two important tasks. Thus, it is both its predictive power and explanatory power, residing in its simple equations, which enable T to compete with its rivals where the search for ever better explanations of the laws governing the natural phenomena becomes imperative. The simpler and more unified T is, given Tr, the greater will be its explanatory power. Beyond this, rational reconstructions of its mathematical, logical, semantical and sociological aspects have remained in focus of the philosophical and scientific discussions in the twentieth century. By a principle of tolerance, advocated by the philosopher Rudolf Carnap[1], it must be recognized that there is a great diversity of rational reconstructions of science possible where philosophers engage in the task of clarifying the nature of scientific theory, theory change and inter-theoretic relations. It is no surprise if philosophers of science belonging to different schools have analyzed the mathematical or logical structure of physical theory differently. Sometimes the principle of tolerance is easily violated when, in contrast with one's own approach, other approaches are rejected as if they were based on a category mistake (Pandit 1995). Consider, e.g. how Carnap (1937, 1938, 1956a,1956b), Quine (1960, 1974, 1981, 1991, 1995) and Kuhn (1962, 1970a, 1970b, 1974, 1977, 1981, 1983, 1989a, 1989b, 1990a, 1990b, 1991a, 1991b) emphasize the linguistic aspects of science, though they do this so differently (Pandit1982, 1991). On the other hand, the semantic conception of the structure of a scientific theory advocated by van Frassen (1989, 221-222) regards the language in which a theory is presented as of no relevance in the discussion of the structure of theories.[2]

If we recognize how complex science and scientific change are in the actual scientific practice, it is to be expected that there will be great differences in how different schools of philosophy view a scien-

---

[1]Carnap, R. (1956a, p. 221); see also Carnap (1937, 1938, 1956b).

[2]For detailed discussions on the semantic and linguistic conceptions of the structure of scientific theory see van Frassen (1980, 1987,1989, 1991), Frederick Suppe (1971, 1972, 1977, 1989), James Mattingly (2005) and Pandit (1971a) respectively. To quote van Frassen (1989, 222): "Of course, to present a theory, we must present it in and by language. That is a trivial point .... But in the discussion of the structure of theories it can largely be ignored". This implies that the structure of classical particle mechanics is not affected by which formulation it is given, a Lagrangian or a Hamiltonian (van Fraassen 1989, 82), simply because it cannot be identified with its linguistic formulation. Articulating this view, Frederick Suppe (1989, 82) points out that theories "are extralinguistic entities which are referred to and described by their various linguistic formulations. This suggests that theories be construed as propounded abstract *structures* serving as models for sets of interpreted sentences that constitute the linguistic formulations".

tific theory. Within physics itself, it is remarkable how in the middle of the nineteenth century the classical theory of the electromagnetic field, once emancipated from mechanics, led the physicists themselves - notably through the work of Helmholz, Hertz, Poincare (and later Weyl) - to view physical theories as symbolic constructions (Dosch et al. 2005b, pp. 41-52). In the later developments, among philosophers it was Ernst Cassirer (1923,1937,1953) who developed this view elaborately as the philosophy of symbolic forms (Dosch 1990). Cassirer's philosophy of symbolic forms is remarkable for its innovative use of the concept of "symbol and its theoretical function to overcome the problems of epistemological dualism". Thus, it is celebrated for turning epistemology around by making the "constitution of sense as the condition of the possibility of knowledge" (Schmitz-Rigal 2002, 2003).

In the twentieth century, most of the schools of philosophy of science using the tools of logical analysis have seen their task mainly as a work of clarification of the logic of science, covering the formal, mainly syntactical and semantical, aspects of the language of science. Consider the following question: How can a scientist rationally justify his acceptance or rejection of a theory T, taking into account not only the relevant body of experimental evidence but also the potential rivals of the theory? As an example of the task of logical clarification, this question concerning scientific rationality contains in its formulation its own answer. For a number of reasons which would become clear below, it is erroneous to assume, as is normally done, as if this is the only major question to be addressed in the task of articulating scientific rationality. Logical empiricists of the Vienna Circle as well as their adversaries in the twentieth century philosophy of science scenario did exactly that, as if it was enough to caricature scientists by asking the following question: How the scientists would accept or reject T, given the relevant experimental evidence E?

In addressing the task of articulation of scientific rationality, taking a static view from outside science has been the norm among philosophers. In particular, in carrying out this very task, philosophers of science have not hesitated in bringing external criteria to bear upon the important task of articulating the aim of science. Karl Popper's approach can be regarded as a good example (Popper1983, xix-xxx). Endless debates about scientific realism and anti-realism are quite noteworthy in this context. Where continuity in the development of science is recognized by the philosophers of science, it is invariably taken to imply the formal requirement that a new theory T2 should reproduce all the successes of the theory, say T1, which is being superseded in its field both in respect of its experimental results and its conceptual framework. Thus, T2 will have to explain all experimental data at least as well as T1. Besides this, T2 will have to correctly predict a new phenomenon, if only to rule out the theories that involve *ad hoc* assumptions. The continuity and rationality of scientific progress implied by this criterion has, however, been challenged by some philosophers and historians of science.[3]

On the contrary, we believe that the challenging task of articulation of scientific rationality can be fulfilled better by looking for the scientific reasoning behind theory change by theory development and theory unification in the particular sciences which one is interested in. To this end, particularly, it is the active frontiers of science which offer the most interesting construction sites where new research questions and novel, if speculative, theories keep appearing on the horizon. No doubt, theory change is so crucial to the growth of scientific knowledge that, following the analytic tradition, one is easily tempted to make trivially true and over-simple statements as follows:

> Given two or more complete theories that compete with one another in their field, the scientist will rationally choose the one which is simpler and which performs far better than its rivals in its field in the sense that it fulfils the aim of science more adequately than they do.

But theory change is not only about how one theory, say T2, replaces another theory, say T1, if it is simpler and if it performs better the tasks of explanation and prediction than its rival does in its field. As we have just hinted at, this philosophical picture is not only superficial but it leaves out the scientific reasoning behind theory change by problem determination and theory development. It is not possible to understand the complexities of theory change without understanding the scientific reasoning based on dynamic core-context building, which triggers developments leading to problem determination and theory development at the active frontiers of science.

In the present study, we want to give an account of theory change by theory development in physics which incorporates those aspects that are either alien or inaccessible to the traditional analytical tools

---

[3]See Thomas Kuhn (1962).

of logical analysis. We prefer to look at science from a dynamical rather than from a static and purely logical perspective. The dynamic approach not only questions but exposes to criticism the philosopher's sharp division of science-studies between the context of justification and the context of discovery. Within this divide, science in the context of discovery has been normally dismissed as not being the proper subject of philosophy of science but rather of sociology or even psychology of science. Also in our view, the contextual relevance of questions relating to the institutions of science and science policy, which may be appropriately classified under "science in context", cannot be denied. We argue, however, that for a sound philosophy of science the problems of what might be called the *context in science* are methodologically highly relevant. Admittedly, in our study, we concentrate on physics, the science meta-theoretically investigated by the philosophers of science, but we think that our arguments apply to all other branches of natural science and even to technology.

Theory change by theory development may involve changes and adjustments in one and the same theory which is developed over a period of time, like the Standard Model of particle physics. Or it may involve several theories, as possible candidates for the kind of theory which is being sought, that are competing with one another for the scientist's rational choice. Two kinds of question concerning scientific theory change can be raised: First, *what* is it in science which changes when theory change occurs and scientific knowledge undergoes revolutionary change? And, secondly, how does theory change by theory development occur at its active frontiers? If one chooses to focus on the linguistic aspects of science exclusively, in-depth syntactical and semantical analysis of the language of science will take precedence over other considerations. From this perspective, linguistic analysis can play an important role in studying what is it exactly which changes in theory change. But if we want to go beyond the linguistic perspective, as we intend to do here, a study from the linguistic perspective has little relevance to the question of the dynamic complexities of theory change or theory development at the active construction sites of science. In particular, it can throw no light whatever on the question how problem determination is possible in physics or how theory-problem interaction and theory-experiment interface shape its active frontiers.

Core-contextual methodological structuralism (henceforth methodological structuralism), which we want to develop here as our response to this question, recognizes the decisive role the structure of physical theory, at the theory-experiment interface, plays in the dynamics of problem determination and problem proliferation. It is in this context that the resolving power of a theory emerges as a most powerful methodological concept. In their turn, problem determination and problem proliferation in physics are known to trigger developments that lead to theory change and, more importantly, to theory development. Thus, methodological structuralism entails that while it is possible to invent new theories of varying resolving powers at the active frontiers of physics, given its dynamic and resourceful core-context building theories in the background, it is not possible to invent research questions formulating scientific problems. Thus, it answers the question how problem determination is possible in physics by recognizing that there is a fundamental asymmetry between theories and problems in physics in the following sense. Scientific problems, unlike theories, are determined in physics in terms of its dynamic core-context building theories, particularly at the interface between theory and experiment. They are determined by employing most powerful theories, i.e. theories which have very high resolving power, which can pose most probing research questions for investigation and which can open new frontiers of research, just as the Standard Model of particle physics does currently.

More specifically, in Chapter 2, it is argued that the policy of dividing the science-studies between the (logical or formal) context of justification and (the extra-logical) context of discovery prevents the essential study of (i) the methodology of inquiry in physics and (ii) correlated theory development and problem determination, including important correlations between theory change and methodological variances. There are multiple relations between the physical theories, the foundations and active frontiers of physics and the methodology of physics. Understanding these is essential for a sound understanding of the dynamical complexities of invention and discovery in physics. From a perspective from inside physics, a significant continuity in its development becomes discernible, the kind of continuity which is not visible from outside in a logical or formal analysis, which has shaped the twentieth century philosophy of science.

In Chapter 3, the general arguments of Chapter 2 addressing the questions of structural identity and scientific change in physics are illustrated in terms of three different case studies in physics.

Chapter 4 undertakes the task of articulating the methodology of dynamic core-contexts of theory development in physics. It analyses the inner development of physical theory characterised by continuities amidst methodological variances. It especially investigates the disappearance and reappearance of important concepts in different core-contextual relations at the frontiers of theory development. The main laboratory of investigation is the Standard Model of particle physics. On the one hand, this theory is recent enough - its main features were set up in the 1960s and 1970s - not to be obscured by so many purely historical questions, but, on the other hand, it is mature enough to allow for a deep philosophical analysis. One of the most important principles in focus here is the principle of gauge invariance, which, after going through several phases of core-contextual development, emerged as a unifying principle – an unique driving force – for the construction of dynamical theory in particle physics.

In Chapter 5, the methodological consequences of the arguments put forth in Chapters 2, 3 and 4 are reviewed, introducing a formalizing model of the core-context building interfaces across theory development and unification at the frontier of fundamental physical theories. It is shown that, at a deeper level, progress in theory development is a process of unification, a process of building more and more interfaces between different core-contexts (Pandit 2002a, 2002b),[4] at times resulting in an extension of the core-context of a theory. In this process of unification and interactive interface building, the structural simplicity and dynamic complexity of physical theory play an important role (e.g., think of the gauge principle in the Standard Model of particle physics). Continuity of progress can, therefore, only be perceived if the full structural simplicity and dynamical complexity of a physical theory and its formalism are taken into account. Besides the progress alluded to above, there is another mode of progress which takes physics *forwards* without interruption. This is the mode of progress by problem determination, using old and new theories. Problem determination allows physics to widen the dynamic core-context already in place and to create new core-contexts. In the ideal case, they all lead to unifications.

---

[4]The dynamic core-context of a theory, as we introduce and understand this concept in this book, and the "hard-core" of a scientific research programme in the sense of Imre Lakatos (1970, 1971a, 1971b) are different concepts altogether. For a clarification, turn to Chapter Two, sec. 2.1.1 and Pandit (1991, pp. 61-62, 129-132).

In Chapter 6 and 7 we cover some of the necessary physical and philosophical background for advocating methodological structuralism in this study. The Epilogue briefly sums up some of the conclusions we arrive at in this study

Physics builds itself by theory-building on the one hand and by advances in instrumentation and experimental testing of its theories on the other. In certain contexts, where theory-building is far ahead of technological limits of experimental testing of theories, physics builds upon *Gedankenexperimente*, if only as fore-runners of actual experimental testing at an appropriate time in the future when necessary technology is available. It is not surprising that, in situations like this in physics, it is core-context building which plays a predominant role in theory-development. In our study, we are concerned with theory-development in physics as a function of the dynamic core-context building.

Leaving out of account all the rest might give rise to the impression *as if* in our view all other just alluded aspects were less important. This is simply not correct. The same consideration applies to history of physics, or history of its other branches, historiography of physics and history of philosophy of science. All of them can be shown to be important in their own right, particularly if we want to gain insights for purpose of articulating theory-development in physics. But due to its limited scope, it is simply not possible to include them for discussion in our study.

Finally, we hope to take the reader interested in understanding scientific reasoning behind problem and theory development in physics a step forward by articulating and advocating methodological structuralism. To this end, we confine ourselves in this book mainly to the frontier theories of particle physics and to a lesser extent to those of cosmology. We think, however, that the path of articulation proposed here can also be followed in an investigation of applied physics and technology studies in general. This hope is motivated by the strong interface that has built itself up during the past half a century between the advances in "fundamental physics" and those in technology. In particular, the advances at the experimental frontier of physics, making theory-experiment interface not only dynamic but decisive for theory change by theory development, are most noteworthy.

Chapter 2

A CENTURY OF OPPOSITION TO THE METHODOLOGY OF INQUIRY

Aber zu einer vernünftigen Frage gelangt man nur mit Hilfe einer vernünftigen Theorie. Man darf nämlich nicht etwa glauben, dass man über den physikalischen Sinn einer Frage ein Urteil gewinnen kann, ohne überhaupt eine Theorie zu benützen. Vielmehr kommt es häufig genug vor, dass eine gewisse Frage nach der einen Theorie einen physikalischen Sinn hat, nach der andern Theorie nicht, und dass sie daher ihre Bedeutung zugleich mit der Theorie wechselt. Max Planck (1965, 240)

2.1 The Structural Simplicity and Dynamical Complexity of Physical Theory

Responding to the revolutionary developments in the sciences during the first few decades of the 20th century, one of the most leading representatives of logical empiricism Rudolf Carnap (1937, p. 277) stated the aim of philosophical analysis as follows:

"The questions dealt with in any theoretical field ... can be roughly divided into object-questions and logical questions. By object-questions are to be understood those that have to do with the objects of the domain under consideration, such as inquiries regarding their properties and relations. The logical questions, on the other hand, do not refer directly to the objects, but to sentences, terms, theories, and so on, which themselves refer to the objects. (Logical questions may be concerned either with the meaning and content of the sentences, terms, etc., or only with the form of these ...".

Articulating this aim further, in accordance with the conception of philosophy newly developed by him and the members of the Vienna Circle, Carnap (1937, p. 279) added that

> "...the logic of science takes the place of the inextricable tangle of problems which is known as philosophy".

Around the same time, parallel to the logical empiricist response to new scientific developments, there were other serious attempts in the same direction which focused on reforming philosophy in terms of a new programme of logical analysis. For example, focusing on science and scientific discovery, Karl R. Popper (1934; 1959, p. 27) declared that

> "A scientist, whether a theorist or experimenter, puts forward statements, or systems of statements, and tests them step by step. In the field of the empirical sciences, more particularly, he constructs hypotheses, or systems of theories, and tests them against experience by observation and experiment. I suggest that it is the task of the logic of scientific discovery, or the logic of knowledge, to give a logical analysis of this procedure; that is, to analyse the method of the empirical sciences".

During the later half of the 20th century, science has advanced far beyond our expectations not only at its theoretical and the experimental frontiers but also at the methodological level of theory-development. The question arises how far these developments are reflected in the philosophy of science, which has set itself the limited aim of articulating the nature of scientific rationality in its various aspects. It is not an easy task to address this question in all its complexity, since answering it will involve an attempt to go into the detailed history of science and different philosophies of science themselves. In this context, it may also be necessary to look for, and characterize, the similarities and differences among the dominant traditions of philosophy of science of the twentieth century, in which philosophers have tried in divergent ways to articulate the nature of scientific rationality.

It is not our aim here to try to answer this question in all its complexity. It should suffice for our purposes to recall how the rise of the analytic tradition at the beginning of the twentieth century set the scene for logical analysis of language, including the language of science. The analytic tradition shaped and dominated the philosophical thinking within the major schools of philosophy in the past century. Logical empiricism led by Moritz Schlick (1949), Rudolf Carnap (1937, 1956a) and other members of the Vienna Circle in the 1920s and critical rationalism led by Karl R. Popper (1934) in the 1930s and after can be regarded as pioneers in this context. Thus, the answer to our question above can be found in the critical analysis of the *logical syntax of language*, as the logical empiricists and other representatives of the analytic tradition have carried it out with the aim of clarifying the logic of science. Even at the beginning of the twenty-first century, the followers of this school remain exclusively focused on this aim.

Within this tradition, the logical structure and presuppositions of science invite questions which include those that concern the logic of confirmation or validation of scientific theories - questions of how scientists might deal with theories in order to test them for rational choice or acceptance (Pandit 1971a). Taking scientific theories as given and as complete in all their linguistic and formal aspects, wherever methodology of science figures as a subject of discussion, it is dominated by the logic of science. For example, this is true of Karl R. Popper's attempt (1934, 1959, 1972a, 1972b) to find a solution to the problem of demarcation which he quite admiringly calls Kant's problem:

> Which criterion "would enable us to distinguish between the empirical sciences on the one hand, and mathematics and logic as well as 'metaphysical systems' on the other (Popper 1959: 34, 1972a: 93)"?

In dealing with this problem, it is the *logical form* of the universal statements of science which is most decisive for Popper (1959: 40-42), both in the task of formulating the criterion in terms of his principle of falsifiability and in building his methodology of theory-testing and theory-change. Invariably, logical contradiction between universal statements and "basic statements" (of observation) is assigned by him a fundamental methodological role (Pandit, 1971a, 1971b). The same is true of the logical empiricist (with Carnap 1937, 1956a, Herbert Feigl 1956, Moritz Schlick, C. G. Hempel 1965 et al. as chief representatives) and the syntactic and semantic theories of science and growth of knowledge (Suppe 1977).[1] Besides these, we come across various kinds of variations on the same theme. If we want to familiarize ourselves with any further developments that might have taken place in the last century, Thomas Kuhn (1970, 1977) is a good example.

As physical theory developed from Hermann von Helmholtz, Ernst Mach, Henri Poincare, Heinrich Hertz, Max Planck and Pierre Duhem to Albert Einstein, Hermann Weyl, Niels Bohr and Werner Heisenberg, philosophy increasingly emerged as an important context of scientific discovery. However, during the second half of the 20th century and at the beginning of the 21st century, the connection between philosophy and physics must be closely interrogated. Physicists looking for guidance have probed the challenging "unreasonable ineffectiveness" of philosophy in physics, viewing their connection with a sense of skepticism the more and more it has become problematic, seemingly even irrelevant. This is not surprising, since theory development within physics has been neglected in the philosophy of physics. Today it is all the more important to interrogate our understanding of how physics has made progress by theory-development and by relentlessly trying to bring the laws of nature within a single unified framework. What are the guiding principles at work here? Is it possible to reconnect physics, which is itself in search of a theory beyond the Standard Model of elementary particles, with the philosophy of physics in some significant sense?

Today, for many of us the task of connecting and re-connecting physics with the methodology of physics may appear to be a problem as well as a great challenge. This is particularly so, if we recall with what ease the scientists and mathematicians Hermann Helmholtz, Hermann Weyl, Henri Poincare and others had recognized those problems which were regarded as common between philosophy and science. The dilemma can be explained only in terms of the *great divide* which was invented and shared by most schools of the twentieth century philosophy of science. We are here referring to the great divide between the scientific theory itself, the *intended object of logical analysis*, and philosophy of science as *the meta-theory* within which all analysis is required to be conceptualized and realised. Within this paradigm, it is widely believed that while philosophy must study the logical structure of a scientific theory wherever, in its completeness, it becomes

---

[1]As regards logical empiricism, detailed discussion can be found in (Pandit, 1971a, 1982, 1991).

accessible syntactically and semantically, *it* cannot study its inner development or the context of scientific discovery. Without interfering with science in any way, the *great divide* was sustained mainly by the logical over-simplifications which the leading representatives of the analytic tradition, notably Hans Reichenbach, Rudolf Carnap and Karl R. Popper (1983, xix-xxx.),[2] have employed in order to articulate the logical structure of scientific theory. It was based upon the following assumption:

> Instead of continuing with the old tradition of metaphysical system-building, it is a better option for philosophy of science to study the possibility of the sciences such as physics and astronomy.

According to them, the best way to fulfilling such a task was by analysing not only their linguistic complexities but the logical forms of their statements and theories, using the logical tools of analysis drawn from symbolic logic and set theory. Philosophy and methodology which had descended from *Philosophia Naturalis* now became so reductively identified with the *logic of science*. We shall later discuss some consequences of this development in the section 2.2.

The meta-theoretical approach of the twentieth century philosophical traditions uses sophisticated methods of looking at *science* from a perspective lying far outside of it (see e.g. Pandit 1991: 184-197). Looked at from outside, sciences of physics and astronomy seem to present a picture of great accomplishment, where great theories offer themselves for *logical analysis*. However, greatly fascinated by this picture, as the philosopher and the non-philosopher are, they lose no time in declaring their vocation with regard to scientific knowledge production in physics. In particular, the philosopher of science asks the question: What is the logical structure and function of physical theory as the *end-product* of scientific inquiry, taken syntactically and semantically in all its completeness? Given two or more theories in the same field, what is the nature of their inter-theoretical relations? How can one of them be *reduced* to the second theory? Thus, from

---

[2]While discussing the aim of science, Popper (1983, xix-xxx) tells us: "Since the publication of the *Logik der Forschung* (that is, since 1934) I have tried to start with some suggestion about the aims of scientific activity, and to derive most of what I have to say about the methods of science - including many comments about its history - from this suggestion".

outside, the philosopher takes a static (*synchronic*) view of science and scientific knowledge production.

Consider now the actual scene of scientific development. Today, the active frontiers of theory development in physics and cosmology present a new scenario full of important methodological challenges. For example, one of the methodological challenges here can be best summed up by asking the following question:

> How is it that particle physics can directly contribute to studies of the large structures of the universe, e.g. the rotation of galaxies, galaxy clusters and even larger structures?

The recent developments at the interface building up between physics and cosmology present an unprecedented methodological situation where the traditional philosophical categories of interpretation fail, challenging even our traditional categories of understanding scientific progress. In order to study these developments, it becomes imperative to explore how far it is possible for philosophy of physics to go beyond the standard theory and practice of philosophy of science. Our discussions in this and the following chapters are intended as a step in that direction. It is our goal here to articulate the dynamic complexities of theory change by theory development, trying to re-build the interface between physics and the philosophy of physics, employing actual historical case-studies while going beyond mere meta-theoretical logical analysis in the analytic tradition.

2.1.1 Questions of Existence, Logic and Symbolic Representation

As a leading representative of this very meta-theoretic tradition, the logical empiricist Rudolf Carnap (1938, 42-62) described the philosophy of science as a study not of the actions or beliefs of scientists but of their *results*, namely, science as a corpus of ordered knowledge. Not the psychology of science but the statements asserted by scientists, forming its *results*, were thought to be a subject of philosophical study and analysis. "The task of the theory of science in this sense," wrote Carnap (1938, 42-43), "will be to analyze such statements, study their kinds and relations, and analyze terms as components of those statements and theories as ordered systems of those statements". Thus, the analysis of the linguistic expressions of science was considered as the main task of the *logic of science* (Carnap 1938, 43). Later on, in his essay "Empiricism, Semantics and Ontology", Carnap's (1956a, 205-221) approach assumed considerable sophistication, recognizing complexities within science that were simply ignored in the earlier phases. A scientific theory came to be viewed as a *linguistic framework* in which the scientist wants to speak about a new kind of entities. Where traditional philosophers have raised problems of existence and reality without being able to solve them or stop them from proliferating, Carnap (1956a, 206) now urged to "recognize a fundamental distinction between two kinds of questions concerning the existence or reality of entities". These he called *internal* and *external questions* of existence, respectively. The *internal questions* are questions of the existence of certain entities of the new kind *within* the linguistic framework. External *questions* are questions of the existence or reality *of the system of entities as a whole.* Elaborating on this, Carnap (1956a, 206) wrote: "Internal questions and possible answers to them are formulated with the help of the new forms of expressions. The answers may be found either by purely logical methods or by empirical methods, depending upon whether the framework is a logical or a factual one. An external question is of a problematic character which is in need of closer examination". Re-visiting his meta-theoretical approach to the problem of understanding and analysing scientific theory, Carnap (1956a, 221) reformulated the basic principle of logical empiricism as follows:

> "To decree dogmatic prohibitions of certain linguistic forms instead of testing them by their success or failure in practical use, is worse than futile; it is positively harmful because it may obstruct scientific progress. Let us learn from the lessons of history. Let us grant to those who work in any special field of investigation the freedom to use any form of expression which seems useful to them; the work in the field will sooner or later lead to the elimination of those forms which have no useful function. Let us be cautious in making assertions and critical in examining them, but tolerant in permitting linguistic forms".

In order to illustrate and examine the distinction proposed by Carnap, let us consider how we might extend Carnap's meta-theoretical approach to the task of analysing the language of physical theory. As

21

our example, we may consider the Standard Model of particle physics. Let us assume that it introduces new language forms into the language of physics ranging over a whole hierarchy of the strong, the electromagnetic and the weak interactions, along with a new system of entities. The entities in question range over the realm of sub-nuclear particles up to the current experimental limit of energy where one can probe distances down to 10-16 cm. According to Carnap's rational reconstruction of science (1956a, 205-221), it is possible to formulate certain questions as theoretical questions with the help of these new language forms, which are to be regarded as *internal questions*. For example, one may regard as *internal questions* those questions which the particle physicists have been pre-occupied with for many years. These include the question whether the Higgs boson exists.

Internal questions, such as these, must be, however, distinguished from those questions which arise from outside this theoretical framework *but* which lack a proper formulation. According to Carnap, such questions are to be regarded as "external questions". They have traditionally been viewed as "questions concerning the existence or reality of the total system of the new entities". In a strict Carnapean sense, "external questions" have to be understood as the metaphysically neutral re-incarnations of old metaphysical questions asked by philosophers in the past. In Carnap's words:

> "Many philosophers regard a question of this kind as an ontological question which must be raised and answered *before* the introduction of the new language forms... In contrast to this view, we take the position that the introduction of the new ways of speaking does not need any theoretical justification because it does not imply any assertion of reality... An alleged statement of the reality of the system of entities is a pseudo-statement without cognitive content. To be sure, we have to face at this point an important question; but it is a practical, not a theoretical question; it is the question of whether or not to accept the new linguistic forms. The acceptance cannot be judged as being either true or false because it is not an assertion".

On the contrary, in Carnap's view, introduction of a new framework in physics is a matter of practical decision. A practical decision can turn out to be fruitful and conducive to the aim for which the language is intended. As a consequence, all philosophical questions that have the appearance of theoretical questions must be rejected and dissolved as cognitively meaningless. Only those questions which survive after proper scrutiny as *re-incarnations* of practical questions in the above sense are to be admitted as worth pursuing. In a nutshell, most if not all traditional metaphysical and ontological questions of existence and reality are ruled out as cognitively meaningless.

But now one might ask: Whether there still remains anything which could be called philosophy of science? Does not something of the philosophy of science still survive, if it is not allowed, following Carnap, to raise theoretical questions of existence and reality but only permitted to formulate *proposals* for practical choice from among a diversity of *linguistic frameworks*? We think that Carnap sees the external questions of existence in a context which is too narrow. Evidently, he accepts as an alternative to his positivistic point of view only an entity-realism. The purely positivistic view has not only been criticized by many eminent physicists (Planck, Weinberg and others) but has been subjected to a decisive philosophical critique by Ernst Cassirer (1937). Taken as a constructive symbolic form (Dosch 2005b), in science the question of existence cannot be separated from the formalism. Before Cassirer, the conception of science as a symbolic (and hence not excessively logical) construction has also been shared by such eminent scientists as Hertz, Helmholtz, Poincare and Weyl.

For example, consider the role of quarks in the Standard Model of particle physics. Quarks as isolated particles have not been found where experimental detection is concerned, not-withstanding their fundamental role. The problem of quark-confinement is quite well-known in particle physics. Since particles are particles only if by the conventional wisdom of experimental detection they are known to exist as particles *and* since quarks cannot be found as isolated particles in this sense, one must consider the possibility that they are not particles proper. This raises the issue how are we to deal with the question "Whether quarks exist?" which is a question of existence that Carnap calls an internal theoretical question. If quarks are not regarded as particles, how are we to deal with the above question? Physicists consider quarks as "essential symbols in the description of sub-nuclear phenomena" (Dosch 2008, p.250). Since the Standard Model of particle physics, in which quark fields occur indispensably, has many observed consequences (Dosch 2008, p. 250), one might quite reasonably conclude that quark fields exist in a sense which goes beyond the conventional experimental wisdom. Quite significantly, this implies that the concepts of a physical theory cannot be separated from the framework of the theory in the manner in which Carnap's above distinction between internal and external questions of existence suggests. Let us elaborate this point a little further.

Since the concepts of physics can be regarded as symbolic constructions, they play their significant role only within a complex formalism. Since no formalism is itself unique, there is no unique language of physical theory. In this sense, it is possible to agree with Carnap's linguistic approach to a limited extent only when taken together with his principle of tolerance (1956a, 221), which permits the construction of alternative linguistic frameworks. However, we want here to express our partial disagreement with Carnap. It is not possible to accept his claim that while the acceptance or rejection of a linguistic framework as a whole is just a matter of practical decision regarding the choice of the structure of language, *the debate* about the existence of entities *within* this framework is strictly theoretical and scientific in its character (Pandit 1991, 95-106, 1971a). It can be very well argued that it is implausible to identify the theoretical framework, within which the physicist must work, *with* a language or with a linguistic framework. Consider the *seemingly* two different kinds of concerns, one involving the physicist and his engagement with theoretical questions *within* the linguistic framework L underlying his theory T and the other involving the philosopher and his engagement with practical questions concerning L as a possible framework for T. We believe that they cannot be put in separate categories. They cannot be sharply separated from each other as Carnap manages to do under the linguistic compulsions of his meta-theory.

If we want to take another look at physical theory while trying to move beyond *the meta-theoretical tradition* of Carnap and his contemporaries, then which is the way forward? The answer is as follows: The challenging task ahead is not only to expose the shortcomings of this tradition but to *re-connect* physics with the philosophy and methodology of physics in a way which does not land us in Carnap type *methodological conventionalism* with regard to the language of science (Pandit 1991, 95-150). Not only is the methodology of physics itself trivialized by *methodological conventionalism* but one begins doubting seriously whether the theoretical framework of physics is really what philosophers of physics working in the meta-theoretic analytic tradition take it to be.

What is worse is the following consequence of Carnap's meta-theoretic approach. Combined with his *methodological conventionalism*, it reduces the task of the methodology of physics to the logical analysis of physical theories taken as linguistic frameworks. Acceptance or rejection of physical theories, as linguistic frameworks, becomes then a matter of practical decision. Once chosen or accepted, the physicists can, within these frameworks, talk about the newly introduced entities, their properties and interactions. However, it will not be surprising at all, if they completely lose sight of physical theories in their fundamental aspects. In particular, we are here concerned with physical theory in those aspects where it builds an active frontier of new probing research questions that take the physicist beyond the framework of the existing theory (Pandit 1991). The search for ever better alternatives to existing theories in physics cannot be explained by Carnap's methodological conventionalism simply because it makes the construction and choice of theories a matter of *purely* practical decision. Those physicists who are engaged in this search do not only have to make a practical decision, be it theory-development, theory-testing or theory-choice. But they follow also other lines of research to be discussed in this book later. If Carnap's discussion of linguistic frameworks has to have any relevance to our understanding of physical theories, then the construction and choice of a linguistic framework cannot be just a matter of practical decision only. At this point precisely, one may start doubting the real intentions and strategies behind his methodological conventionalism. One may start wondering over the question: What happened to the original subject of analysis and appraisal, viz. physical theory? Is it still the same *subject of discussion* or has it radically changed beyond recognition (Pandit 1982, 1991)? After all, one might seriously doubt whether physical theories and linguistic frameworks are identical *subjects of discussion* at all? In this very context, consider yet another shortcoming of the purely meta-theoretic/meta-linguistic approach of Carnap: How are we to account for the existence of different languages which are sometimes necessary to explain several aspects of the phenomena treated by *the same theory* (Dosch et al 2005b, 9-17)?

Now in the light of the scientific developments of the past century which have shaped theory development in physics (some of which will be discussed in Chapters 3-5), it is most appropriate to pursue the question: Whether it really makes any sense to suggest, following Carnap, that *the questions of existence* within physics and those of *the theoretical framework* within which answers to such questions must be explored can be separated sharply enough from one another in order to put them in different *categories*? Does it make sense, for example, to treat the question whether Higgs boson exists as a theoretical question within the linguistic or conceptual framework of the Standard Model of particle physics, while treating the questions concerning its framework as a whole as if these were questions of only practical choice as to which framework to adopt? We can argue against Carnap's distinction as follows: If the possible answers to the latter are taken to be governed by methodological conventionalism in Carnap's sense, there is no reason why a more thoroughgoing methodological conventionalism should not govern possible answers to the internal questions. It would be both interesting and instructive to explore how the physicist who is confronted with problems of invention and discovery at the active frontiers of physics might respond to this. The development of particle physics through the various phases presents an instructive methodological model of theory development where questions of existence of the individual particles and those of the theoretical framework for pursuing them turn out to be of the same category. No particle physicist would separate these questions under different categories. The question which of the particles that are being dealt with or postulated exists is as much a theoretical question in particle physics as the question: Within which theoretical framework should we pursue such a question? In Chapter 3 (secs. 3.1.2 and 3.1.3), we shall, therefore, briefly begin with an important case-study in theory-development, as a counter-example in support of the point we have just sketched against Carnap's distinction.

Here it is important to ask the question: What were the motivations behind the distinction which Carnap drew between (i) the scientific questions of existence that are *internal* to the science of physics and (ii) the unscientific practical questions that are *external* to it? The former are regarded by him as internal questions not only because they arise within the framework of a whole physical theory but also because possible answers to them are required to be formulated as assertions which are possibly true or false. Similarly, the latter are regarded as external questions not only because they concern the frameworks of whole physical theories - or the system of entities as a whole which they introduce - but also because answers to them are required to be put forward as proposals to accept or reject a certain framework as a matter of convention or practical decision only. Two things are noteworthy here. First, as a philosopher of science Carnap's motivations in drawing his distinction were manifold. One of these had to do with his strategy to arrive, if possible, at the re-incarnations of some of the traditional philosophical theories and problems, showing how they have to do with the structure of language rather than with the metaphysics of "reality" or "the world" as claimed by their authors. From this step, it would have been easy to take the next step envisaged by Carnap, demanding that philosophy of science be invested with the task of proposing language forms for science instead of just clarifying the logic of the language of science. Many of Carnap's works are dedicated to these themes explicitly (see e.g., Carnap 1937, 1938, 1956a). Secondly, Carnap's distinction is itself based upon an over-simple picture of science such as physics. In this picture, it is not possible to account for theory-development in physics. The question is if physicists succeeded in establishing that the Higgs boson is not found, how would physics be guided in finding the ways and means of modifying the Standard model of particle physics suitably? No doubt, it would be guided by the negative feedback from such a negative discovery. But more importantly, as we shall be arguing later in this chapter, it would be guided by the core-context in physics. If we accept Carnap's distinction, none of theses possibilities should make sense.

To put the point differently, in order to deal with the structural simplicity and dynamical complexity of physical theory, it is necessary to go beyond Carnap's distinction. As a first step, it is necessary to draw a three-fold distinction between: (1) the physical theory T, its fundamental principles, its formalism, and its relation to observables; (2) the novel predictions which can be derived from T, involving the questions of existence of the newly predicted entities; and (3) the probing research questions which T asks, posing new problems at the frontier of development and enhancing the possibility of unifying formerly separated fields. The last point has far-reaching implications. It tells us how important the core-context in physics is. The core-context in physics plays an important role in its inner development, especially in problem-determination and theory-development. Without this three-fold distinction, which we shall articulate in sect. 2.3. and develop in later chapters particularly Chapter 4, it is neither possible to understand theory development in physics nor to explain the appearance and disappearance of important concepts in different core-contexts in physics.

2.1.2 Asymmetry between Problems and Theories in physics

Puzzles and puzzle-solving on the one hand and problems and problem-solving on the other have dominated the conceptualizations of scientific change and scientific progress that have been attempted within the twentieth century philosophy of science. Thus, Karl Popper (1972a, 1972b) has held the view that science always starts with problems and progresses only when the theory which is invented to solve them is faced with difficulties including contradictions. Unlike Popper, Thomas Kuhn (1970, 1977 and 1993) has argued that the rational scientific activity consists mostly in the elimination of puzzles within a paradigm or a disciplinary matrix. In his view, normal (= rational) scientific activity is nothing but puzzle-solving activity by the scientific community. It is always carried out within the reigning paradigm which defines the standards of scientific practice at a particular time. Kuhn's categories of normal and revolutionary science on the one hand and paradigms/disciplinary matrices on the other entail punctuating discontinuities within a highly fragmented science (Pandit 1982, 1991; Fuller 2000, Maxwell 2005). This calls for a detailed discussion, which we will take up in sects. 2.2.3, 2.2.4. and 2.2.5.

Again, there are other philosophers, e.g. Larry Laudan (1977), who suggest that the rational scientific activity aims at theories which show a maximum of problem-solving effectiveness. Invariably, all of them share the view that theories are meant to eliminate puzzles or solve problems.

In this kind of scenario, one might ask how relevant is the notion of a "hard-core" which is central to Imre Lakatos's (1970, 1971a, 1971b) philosophy of *scientific research programmes*. According to Lakatos (1971a, 99), a scientific research programme consists of the following components: (i) "a conventionally accepted (and thus by provisional decision 'irrefutable') 'hard core' "; (ii) "a 'positive heuristic' which defines problems, outlines the construction of a belt of auxiliary hypotheses, foresees anomalies and turns them victoriously into examples, all according to a pre-conceived plan"; and (iii) a 'negative heuristic'. The last component consists of principles which (a) protect the "hard core" from experimental refutation and (b) rule out radically different sorts of explanatory attempts.

Taken in the context of a Lakatosian scientific research programme, the notion of a "hard core" has to be understood as the main structural and functional unit of such a programme (Pandit 1991:129). Yet, by methodological convention, it is its most protected component. Is this to be taken as an improvement over Popper's methodological conventionalism (Pandit 1991)? In a sense, it clearly is an improvement in so far as it extends methodological conventionalism from Popper's "basic statements" (= potential falsifiers of theories in Popper's sense) to the "hard core" itself (= falsifiable theories in Popper's sense which, according to him, have potential falsifiers). On the other hand, there are objections against it. First, in anaylzing a physical theory in terms of a scientific research programme, while moving closer to the position of Kuhn (Pandit 1982, 1991), Lakatos is still working within the analytic tradition of Popper, Carnap and others. We have seen in our discussion above how this tradition divides science between the context of justification and the context of discovery. Secondly, an important question arises as to which component in a scientific research programme is fundamental or relevant to the structural identity of science. Given a scientific research programme, where can we look for the appropriate structure to which we could attribute the core-context-building principle of the resolving power of a physical theory (Pandit 1991: 131)? We believe that no positive answer to this question is available within a Lakatos-type scientific research programme. It should be noted here that the notion of a "hard-core" as used by Lakatos should not be confused with the concept of dynamic core-context of development in our sense (Pandit 2002a, 2002b; for detailed criticism of Lakatos's doctrine turn to Pandit (1991, pp. 55-63).

It must be pointed out that there is something important which is found missing in this entire scenario. What is missing is a clearly drawn distinction between puzzles and problems in science. For various reasons, the need for such a distinction may not have been felt by the members of the philosophy of science community. Think of the strong assumptions which they employ in their conceptualizations while they are engaged with the main tasks of their meta-theory, which looks at science from a perspective taken from *outside* science. One of these assumptions says that the task of a philosopher is just to analyse the *logical structure* of scientific theory, particularly in its role as explanation of already well-established laws of nature in its domain of applicability. Such analysis is required to be extendable from *the structure* of the theory itself to the structure of its logical relationships with other theories and with relevant bodies of experimental evidence. In the process of such analysis, it is more likely that the philosopher himself lands in conceptual or logical as well as semantic puzzles of meaning and meaninglessness. No surprise, elimination of puzzles became for logical empiricists and other schools of philosophical analysis in the twentieth century one of the main motivations for research. Their main aim was to *unveil* the ultimate logical structure of a language perfect enough for purposes of articulating scientific theories that were thought to be complete in every respect except in their puzzling linguistic appearances. For Russell and Carnap the ordinary natural languages presented themselves as the most simple and readily available examples of linguistic frameworks with puzzling and disguising appearances. Their commitment to philosophy as *meta-theory* and to science as an object of logical analysis in terms of the categories which were strictly dictated by the meta-theory sharply *disconnects and separates* science from philosophy. It is this which brings into philosophy as well as into science linguistic or conceptual puzzles as *a category of analysis* in its own right.

Thus, analysing a physical theory is taken to be a kind of puzzle-solving by the divergent schools within the meta-theoretical tradition of twentieth century philosophy of science. Depending upon which perspective one chooses to adopt from *outside* while focusing on its structure and in articulating its relationships, it will appear differently to experts belonging to divergent schools. It is, therefore, necessary to draw a clear distinction between puzzles and problems, if only to keep certain kinds of confusions from creeping in the discussions, particularly if one wants to look at science from *within*. The question arises whether doing so does not entail a departure from the meta-theoretical tradition of twentieth century philosophy of science. In that tradition, it is normal for one school of analysis to accuse the other of committing a "category mistake" when their conceptualizations don't match, or when they tend to conflict, with one another (Moulines 1995, 301, Popper 1972a, 71). This shows how within the same meta-theoretical tradition, different schools not only fight over concepts and methods of analysis but also over the aims of analysis. As their conceptualizations tend to diverge radically, one loses the sense of alternatives completely, some of which could be better than others under certain conditions. One of the best examples can be found in Stegmüller's set-theoretic structuralism (1979), the so-called non-statement view of science (Pandit 1991). Stegmüller and his followers reconstruct a physical theory radically so differently that it has the following consequence: It is a "category mistake" to talk of *statements* in the context of physics and to ask under which conditions they are meaningful, falsifiable or possibly true or false. No surprise, one of the charges Stegmüller school has levelled against Popper is that in formulating his falsifiability criterion of demarcation he has committed a "category mistake" (Moulines 1995, 301, Popper 1972a, 71). It is another way of acknowledging that philosophical puzzles or pseudo problems about science arise when a philosopher confuses the category of discourse which is both relevant and appropriate to understanding science *with* a category of discourse which is neither relevant nor appropriate. In their set-theoretical structuralist approach, therefore, the problem of demarcation, understood as presupposing a statement-view of science, is a pseudo-problem. We may find other examples in this context, e.g., Kuhn's doctrine of incommensurability, which they will treat in the same manner. We shall discuss this doctrine in sect. 2.2.3. For Stegmueller and his followers who believe in the non-statement view of science, such problems should not arise at all, these being pseudo-problems or puzzles.

What is worse is the following consequence: Irrespective of many differences among different schools of philosophy, where all the attention should have been on the subject of scientific appraisal, e.g., physical theory, one finds that it has already shifted somewhere else (see Pandit 1982, 1991).

No doubt, in the *Tractatus Logico-Philosophicus* (1922), Wittgenstein had ruled out ordinary natural language as useless for purpose of philosophical analysis, being not only most trouble-some but full of puzzling appearances of its hidden structures. He expressed this in his declaration: *language disguises thought*. The puzzles, or the puzzling appearances of language, could be accounted for only by distinguishing two things: The deep structures, called *logical forms*, which are hidden from our view, and the surface grammatical structures that only disguise *thought*. It is well-known that Popper and Wittgenstein had once a confrontation over the latter's suggestion that in philosophy there were only puzzles but no serious and genuine problems (Popper 1972a, 66-96) to solve. Expressing his disagreement with Wittgenstein, Popper even challenged him, pointing out concrete examples of genuine philosophical problems. Popper also criticized the logical empiricist view that the task of philosophy was to analyse the meanings of words belonging to natural or scientific language. He argued that analysis of the meanings of words in natural or scientific language could not be considered under any of those serious problems which may be regarded as genuine philosophical problems. In particular, in this very context, he rejected the problem of formulating a criterion of meaningfulness as a *pseudo problem*, which was accorded highest priority by the logical empiricists. Thus, he differed sharply from all philosophers who were interested either in meaning-analysis or in analysis of beliefs. For logical empiricists, on the other hand, most serious philosophical problems are posed by the meanings of words and sentences within a given language. It is interesting to note how in his *Philosophische Untersuchungen* (1953) the later Wittgenstein's thinking took an altogether different direction, further enriching the diversity of views within the meta-theoretical tradition. Here Wittgenstein saw the greatest puzzle in the fact that there are philosophical puzzles about ordinary natural language at all. As regards the family of ordinary natural languages, and the different language games we play with them in different contexts, he now held the view that everything is in perfect order and open to our view. If there are still philosophical puzzles, it is simply because *philosophical thought itself disguises language*. In other words, philosophical problems, which are now seen as mere puzzles, arise because of our misunderstandings of the way language functions. From this kind of scenario, we may conclude that there is really no agreement among these philosophers even on fundamental notions of what should count as *pseudo-problems, puzzles and genuine problems*, whether in science or in philosophy.

The confusion between problems and puzzles is especially widespread among philosophers. In philosophy, if one can invent a conceptual or linguistic puzzle around a troublesome concept (think of the concepts of "meaning", "truth", "reality" and "existence") and then find a way to dissolve it, it is considered a great philosophical achievement. No surprise, there are philosophers of science who believe that a problem in science disappears as soon as it has been solved. We may call this view *the disappearance view* of problems. This is the kind of view which is strongly dictated by the meta-theoretical tradition we have discussed above (sects. 2.1. and 2.1.1). As we will notice in what follows, it is particularly this view which is so clearly indicative of the confusion between problems and puzzles. It is based on the un-stated metaphysical assumption that the world, which science seeks to describe, is really *simple*. Thus, every solved problem takes us one step ahead towards the intelligibility of the essentially simple world. On the other hand, there are others, among them scientists and philosophers, who believe that problems never disappear completely from science, there being an inverse relationship between the complexity of the world and the simple solutions to scientific problems (Pandit 1991; Maxwell 2012a, 2012b). That is why the old problems, even if thought to be solved once, reappear in some reproblematized form or other (Pandit 1989). The physical world must be complex enough, if it is true that problems do not disappear completely from science once they are taken to be solved. One way of demonstrating this may be to remind ourselves how the simple equations in physics have to cope with complex solutions: the solutions share neither the simplicity of the problems nor the simplicity of the equations.

In physics, on the other hand, puzzle-solving is respectable only in the larger context of a theory (for a case-study, turn to sect. 3.1.3). Thus, a puzzle may arise in the context of theory development taken together with experimental observation. The existence of the puzzle depends on the type of activity the scientist is engaged in before he is puzzled, *as it were* landing in a problem-situation. Appearance of puzzles and their dissolution can occur only in a complex context in which the scientific community is not able, using its skills, to find its way around the "puzzling data". Thus, if the scientific community is working with a theory and its applications, it may come across data which are "puzzling" in the sense that they are neither among the theory's potential applications nor among its potential consequences. They call for their explanation which is not immediately forthcoming or available. However, once dissolved, puzzles stand eliminated. All explanation in science is a movement *backwards* from a known puzzle to puzzle-solving - to the elimination of the puzzle, as Kuhn calls it. A puzzle may have one and only one solution within the scientific community's reigning paradigm in Kuhn's sense. On the other hand, *a problem may have as many possible solutions as its many possible histories. In no case, can a problem have a solution in terms of those very concepts with which it has been formulated.*

In the beginning, it may not be always clear weather a problem is only a puzzle or something deeper than that. In the history of physics, seemingly minor puzzles have sometimes turned out to be major problems which lead to a new theory. Consider, for example, Mercury's perihelion movement around the sun. To an astronomer at the end of the nineteenth century it may have been just a puzzle to be solved within the Newtonian theory. As it turned out with Einstein's general theory of relativity, it became possible to deal with it as a *problem* beyond the framework of Newtonian physics. In absence of Einstein's theory, one could have as well over-simplified the situation. One could have simply declared that Newton's theory was falsified, confronted as it was with a contradiction in terms of Mercury's perihelion movement around the sun, violating the predictions of the theory. No doubt, once falsified, the value of a theory becomes debatable. But the situation is not as simple as, if we follow Popper, we might like to think.

But it is true that a science such as physics *moves forwards* only when it is able to determine real problems, using all the background resources of the theory-clusters, and in some cases even violating the accepted rules of method and strategy (Pandit 1982). If this view is correct, it takes us far beyond Kuhn's puzzle-solving normal science. Objectively speaking, real physics is not to be found there at all where a theory proposes to nature how we might eliminate a puzzle/anomaly in Kuhn's sense. On the contrary, it is to be found there where a theory-cluster poses new problems to nature by posing new research questions. *Problem determination is a subject of enlightened scientific reasoning in a manner which is different from that of the construction of theories and their mathematical formalisms themselves.* Determination of scientific problems depends on whether a science has theories which are powerful enough to pose probing research questions or to offer answers to them, once already posed. Scientific problems cannot be invented in the manner in which puzzles are invented in philosophy.

What about the scientific theories themselves? The answer is both *yes* and *no*. It is correct to say that creative theorizing which is so much characteristic of physics and astronomy cannot be replaced by an algorithm for building and testing theories. Yet, guided by the core-context of development within physics, a scientist can *invent* the kind of theory which is in demand at the active frontier (Pandit 2002a, 2002b). There is no possibility of *inventing* a theory which is totally free from the physical and mathematical constraints of such a core-context. On the other hand, scientific problems as a category in their own right do not depend on the explanation seeking or puzzle-solving activity of the scientist who may happen to formulate or raise them. If a problem has been solved, this will never rob it of its character as a subject of scientific argument - as something that has been not just inherited but determined by scientific reasoning in the course of the historical development of the given science. Even after a problem has been solved, it will continue to enjoy the status of a problem in the same manner in which a superseded theory does.

2.1.3   The Principle of the Core-Context Building Resolving Power of Physical Theory

In Chapter 4 particularly, our discussion will be focused on the core-context guided progress of contemporary physical theory, with the Standard Model of particle physics and the interface it builds up with cosmology in focus. As we shall argue there, this progress can be seen more and more as a result of the *resolving power* of physical theory, viz., its power to pose new interesting and probing research questions which were not thought of before. While a more detailed account will follow in sect. 2.3., here we want to make some general observations only in order to introduce the principle of the core-context building resolving power of physical theory as part of scientific reasoning which plays a crucial role in theory change by theory development in physics. Towards this end, it is pertinent to ask the following question: How is it that the formalism of a physical theory makes it possible for the scientist to make novel predictions about the world? How does it enable her/him to determine new problems? Both novel predictions and new problems are indispensable to the growth of scientific knowledge. By interrogating physical theory in respect of one of its most fundamental aspects, we are already going beyond the meta-theoretical tradition, our subject of discussion above. Already in (Pandit 1982, 1991: 309-326, 2002a, 2002b), there is a move to depart from this tradition by bringing the resolving power of physical theory, which resides in its formalism, to bear upon theory-development and theory-appraisal in physics. it remains to be seen how significantly this might reconnect physics with the philosophy and methodology of physics in the old, forgotten, tradition of natural philosophy (Maxwell 2010, 2012a, 2012b; Pandit 2007c, 2010a, 2010b).

If we continue to ignore physical theory in this fundamental aspect, we would be thereby ignoring much of the scientific reasoning which makes theory development in physics and astronomy possible. As a further consequence, we cannot explain how the scenario of the nineteenth century physics shaped the future growth of theoretical physics. Nor can we explain the progress made by contemporary physical theory. At a fundamental level too, we cannot appreciate how the active frontier of physical theory builds its interface with cosmology (see Fig. 5.1, sect. 5.1.1, Chapter FIVE). We are here referring to that kind of scientific reasoning which combines in a unique way the theory and its consequences with the theory-experiment interface on the one hand and experimental test-results on the other, enabling the scientist to draw important inferences about the structure of the universe, including its fundamental building blocks. The same kind of reasoning enables the physicist to pose new probing research questions which go beyond the range of existing theory. In this way, it plays a very crucial role at the frontiers of theory development in physics (cf. Kane, G. & Pierce, A. eds.2008; and Pierce, A. 2008, pp. 13-23).

None among the great physicists of the 20th century probably thought of what the logical empiricists and other important philosophers of science such as Karl R. Popper were up to when they ruled out any concern with science if taken in *the context of discovery*. After *a century of opposition* to its study, the question which needs to be answered is whether in doing so they did not throw away the baby along with the bath water. As understood in the most dominant traditions of the twentieth century analytic philosophy of science, the context of discovery lies outside the sciences of physics and astronomy. Paradoxically enough, while ruling it out as a possible subject of philosophical rational reconstruction, the philosophers did not even consider the following possibility: In science, the core-context of development not only builds itself up at the theory-experiment interface but plays a fundamental role in theory development in search for unification in physics. In order to explore this possibility, we would like to pose the question:

> Whether we should not explore the deeper aspects of physical
> theory which have remained unexplored so far? Whether we

should not consider seriously the far-reaching consequences of our argument above (sect. 2.1.1.) that the questions of existence in particle physics are inseparably linked to the questions of not only which theoretical framework to employ but which core-context to assume in order to pursue them? Whether, in doing so, we should not take another look at the Einstein-Heisenberg dialogue on physical theory, which enabled them to arrive at an important context principle which says that in physics "It is the theory which decides what can be observed".

Referred to as the Heisenberg-Einstein context principle (Pandit 2002a, 2002b), Heisenberg and Einstein reconnected, even re-integrated, observation with theory, thereby restoring to physics and astronomy a strong theory-experiment interface which they had been robbed of under the earlier empiricist doctrine. We are here referring to the strong empiricist doctrine that a theory must be founded on observable magnitudes alone. On closer scrutiny, this turns out to be a kind of foundationalism in the tradition of Ernst Mach (see Cohen 1968, pp. 132-170). It requires that in its entire structure a theory must include nothing but observable quantities. As we shall see in a moment, at one time it had an overwhelming influence on both Einstein and Heisenberg.

Recalling his discussions on theory and observation with Einstein as early as 1926 when Heisenberg was to give a talk on quantum mechanics in Berlin, Heisenberg (1984, 428-429) tells us how Einstein "listened to the talk and corrected the view" that the kind of philosophy which was most important in quantum mechanics had to do with the idea of introducing only observable quantities. When they met after the talk, at Einstein's initiative, he confronted Heisenberg with the question: "What was the philosophy underlying your kind of very strange theory? The theory looks nice, but what did you mean by only observable quantities"? To quote Heisenberg (1984, 428-429):

'I told him that I did not believe any more in electronic orbits, in spite of the tracks in a cloud chamber. I felt that one should go back to those quantities which really can be observed and I also felt that this was just the kind of philosophy which he had used in relativity; because he also had abandoned absolute time and introduced only the time of the special coordinate system and so on. Well, he laughed at me and then he said "but you must realize that it is completely wrong". I answered: "but why, is it not true that you have used this philosophy?" "Oh yes", he said, "I may have used it, but still it is nonsense!"

Einstein explained to me that it was really the other way around. He said "whether you can observe a thing or not depends on the theory which you use. It is the theory which decides what can be observed...".'

As their discussion developed further, recalls Heisenberg (1984, 429), Einstein pointed out to him that

"it is really dangerous to say that one should only speak about observable quantities. Because every reasonable theory will, besides all things which one can immediately observe, also give the possibility of observing other things more indirectly. ... In quantum theory it meant, for instance, that when you have quantum mechanics then you cannot only observe frequencies and amplitudes, but for instance, also probability amplitudes, probability waves and so on, and these, of course, are quite different objects".

This statement of Einstein, made in a lively conversation should not be interpreted as if physics were totally free of what is measurable. The most important passage is the one concerning the danger, if one relies *only* on observable parameters. In a symbolic description it is quite natural that there occur symbols that do not represent measurable quantities. The wave function (i.e., state vector) in quantum physics is such a symbol. The fact that it is not directly measurable is the foundation of one of the most fundamental principles in modern physics, namely gauge invariance (see sect. 4.4. of Chapter 4).

The example of this dialogue between Heisenberg and Einstein shows how important it is to probe the deeper aspects of physical theory, e.g. by asking the following questions:

(1) Is the concept of context of discovery rather systematically ambiguous, depending upon whether we are interested in looking at science from outside or from inside?

(2) What kind of consequences can be drawn from our argument above (sect. 2.1.1) that the questions of existence of newly predicted entities in particle physics and the questions of their theoretical framework are not only inseparable but more complex in their relationship? In particular, what are its consequences for connecting theory with observation, or for understanding the theory-experiment interface in physics?

(3) What implications follow from the Heisenberg-Einstein context principle, reconnecting observation with theory in a manner which is quite at variance not just with their earlier convictions or beliefs but with those of the logical empiricists?

In order to answer the question (1) in the affirmative, we want strongly to argue that the traditional idea of *science in context* should not be allowed to blind us to *the principles of context-building* in science that guide scientific change within it as it makes progress across the frontiers of development. But that is exactly what has happened during the past century, owing to the influential doctrines of logical empiricists Rudolf Carnap and Hans Reichenbach, among others, and the critical rationalist Karl R. Popper. We think that it is most important today to distinguish between *science in context*, or what the philosophers call the context of discovery, from *the context in science*. Philosophers and scientists may choose to deal, or not to deal, with the former. But we believe that they cannot afford to ignore the latter, namely *the context of development* in science such as physics. In order to differentiate sharply *the context in science* from the philosophical conception of *the context of discovery* in the tradition of Carnap, Reichenbach and Popper, we propose to designate *it* as the *core-context of development*. This implies recognition of the fact that there are dynamical principles of context-building at work within physics. These principles are discernible particularly in the current physical theories which have matured sufficiently enough, not only creating new frontiers of development but demanding unification of the fundamental laws of nature within a single framework. Quantum field theory is a good example.

Now the question may be asked how this dynamic core-context-building is possible in physics. We may answer this question by considering the proposal that we can understand it as a function of the resolving power of physical theory (Pandit 1982, 1991). This proposal is premised on a distinction between two aspects or roles of a physical theory, the one which deals with explanations of phenomena or laws governing them and the other which poses new probing research questions at the frontier and makes novel predictions, inviting attempts at novel experiments to discover hitherto unknown phenomena of nature. Scientific progress in physics as in other natural sciences involves a movement from one milestone to another. A dynamics of methodological variance, within an overall invariance, is at work in physics as it follows the twin methodology of (a) *working backwards* from the antecedently known problems to their solutions in the form of theories and (b) *working forwards* from theories to problems (Pandit, 1982, 1991, 2002a, 2002b). Triumph in physics is not only to be found in (the *invention* of) the all encompassing theories the physicists are capable of conceiving, it is also to be found in problem-determination, using all its current theories as the core-contexts of development (Pandit 1991, 2002a, 2002b), where the methodology of *working forwards from theories to problems* must complement the methodology of *working backwards from problems to theories*.

This recognition leads us to asking the following question: Whether, besides its strategy and methodology of inquiry (for more on this topic, turn to Chapter 6), physics does not have, amidst all its *dynamical complexity*, some kind of *structural identity*? How can we identify physics or astronomy, or any other natural science, structurally or methodologically, taking into account its successes or achievements as well as its newly opening frontiers full of new problems? It is possible to answer these questions differently, depending upon from which perspective we are looking at these sciences. We propose to offer a tentative answer along the following lines.

At any stage of its development, a science such as physics must have well-identifiable structures which are marked with its present and past successes. More importantly, it must have built into the mathematical formalisms of its theories the future possibilities of development in terms of both new theories and new problems. Only under this condition of structural identity, marked with dynamical complexity, can it claim that the directions of development which its practitioners may follow are not without the guiding principles which are themselves rooted in that structure. The maturity and success of its theories will be finally judged by their power to guide further research and to create new frontiers of discovery. Thus, wherever physics has a well-identifiable frontier of discovery created or triggered by its sufficiently rich fundamental theories, we can refer to the participating theories and the theory-experiment interfaces as its *core-context of development*. What qualifies them for this role is their ability to build up inter-theoretical relations within physics thereby setting up *the core-context of development*. In other words, we want to put forward the following thesis: If physics always has a frontier of discovery which makes it susceptible to scientific change, then, closely correlated with that, it must have *a core-context of development*. The very possibility of a frontier of discovery must be premised upon its successes and achievements which can be held constant in the search for new knowledge. *The core-context of development* in physics holds constant precisely those achievements and interfaces which guide the physicist at the new frontiers of discovery. Whenever there is a breakthrough at the frontier, say by inventing a new theory, the fundamental past achievements should be able to build an interface with the new theory and with the new experiment.

Heinrich Hertz (1962) and P. Duhem (1906, 1914) were among the first to clearly hint at the role the contexts of theory-testing play *within* science. In the context of his experimental and theoretical work on the principles of mechanics, Hertz arrived at a holistic conception of physical theory. He wrote (1962, 197):

> It is true, meanwhile, that each separate formula cannot be specially tested by experience, but only the system as a whole. But practically the same holds good for the system of equations of ordinary dynamics.

In a significant step forwards, Duhem (1906) made the holistic conception a cornerstone of the structure of physical theory. In his own work, he referred to those contexts where clusters of theories/hypotheses participate in the experimental testing of a newly proposed theory. Given a theory under test, the test-results are to be jointly attributed to the participating theories. Crucially, if the predictions fail and the test-results are negative, no single theory can be held responsible. Following this line of reasoning, like Duhem, we are now asking the question whether there do not exist *core-contexts of development* within physics. Whether there do not build up such contexts where present theories, which are already successful and confirmed, participate in the process of inventing the kind of theory which physics at its frontiers, at the theory-experiment interface, can aim at?

There are at least three kinds of possibilities which we must distinguish from one another. *First of all*, in accordance with both our argument above (sect. 2.1.1) and Heisenberg-Einstein context principle (Pandit 2002a, 2002b), we can so reconnect observation with theory and reconstruct theory-experiment interface that physics can be said to aim at those theories which fulfil this principle rather than those which do not do so. To think of *the core-context of development* in physics is, however, to think of those theories which go beyond the Heisenberg-Einstein principle, since it merely states *that it is the theory which decides what can be observed or measured. Secondly, a core-context of development* may emerge in physics when the participating theories are those which provide crucial inputs to the methodology of theory change by theory development and theory unification in physics. They must all participate in the search for a new theory in which the earlier theories would be allowed to live harmoniously and contradiction-free. As a rule, they must be highly successful so that they participate in this methodology. Of course, the participating theories or theory must be successful in other important respects. Take the case of quantum mechanics and the general theory of relativity. As such they have not (yet) been shown to be compatible. The search for a quantum theory of gravity is an active frontier in physics today. If a quantum theory is found which applies to all the four forces of nature, it may be possible for the existing theories to live in it contradiction-free. *Thirdly*, in physics a *core-context of development* emerges when the existing theories acquire sufficient maturity to become the frontier-creating theories. Einstein's general theory of relativity amply fulfils this requirement. As a theory which is sufficiently rich, mature and powerful, it has created a whole new frontier of discovery at the very core of astrophysics (e.g. research on black holes and binary pulsars). Taken in this sense, *the core-context of development* in physics deserves serious scientific and philosophical attention. We think that the future of the philosophy of science will depend upon how we go about the task of understanding the core-context of development in physics and possibly in other sciences.

2.2 The "Unreasonable Ineffectiveness of Philosophy": Methodology and Philosophy of Science

Since the time of Newton physics has consistently followed a methodology which enables it to search for simple and unified dynamical theories. This is expressed in the famous *regulae philosophandi* of Isaac Newton as follows (Rules 1, 2 and 3):

1. To admit not more causes of natural things than those which are true and are sufficient to explain their phenomena.

2. And therefore the same causes have to be assigned to natural effects of the same kind, whenever this is possible.

3. The qualities of bodies which can be neither increased nor decreased and which belong to all the bodies for which experiments are feasible, are taken to be qualities of all bodies

Whereas Rule 1 and Rule 2 can be seen as modifications of Ockham's razor, Rule 3 is especially important: It is directed against the dogmatic rational philosophy of Descartes, who postulated that material bodies are exclusively determined by their extension in space, thereby opening the way to 'experimental', that is empirical, science. At the time of Newton one might have thought of the one and the only one physical theory. The development of physics has revealed a much richer structure of parallel, sometimes competing, sometimes superseding theories. For these theories the modern philosopher of science has formulated a principle of invariance as follows: That each theory asserts that "throughout the range of phenomena, actual and possible", to which it applies, "the same laws govern the way these phenomena evolve in space and time" (Maxwell 2006). By searching for one kind of physical theory instead of some other kind, the physicists themselves have been able to set goals for physics, progressively moving from one kind of theory to another without violating the principle of invariance. Interestingly enough, even the Bohr-Einstein debate and the Heisenberg-Einstein dialogue on quantum mechanics are essentially about the physicist's search for the kind of theory which physics can legitimately and realistically aim at.

In the dynamic core-context (of inquiry) developing within physics itself, what is most surprising is that the mainstream philosophy of science in the twentieth century has remained unaffected by this very important and fascinating aspect of modern physics. As we have shown above, the philosophers of science have been mainly concerned with the logic of science - i.e., with the logical analysis of the language of science. In other words, interested exclusively in the tradition of meta-theoretical analysis, they have been mainly engaged with its achievements, or final results, such as the scientific theories that are being put forth as candidates for confirmation, or acceptance, given the necessary experimental evidences. **Neither the dynamic core-context of development nor the methodology of inquiry within physics has attracted their attention.** The question arises whether we can find out the reasons why this has been so. Whether we can find out what consequences their opposition to the methodology of inquiry has had for our understanding, or investigation, of the multiple roles which a theory can play in science such as physics and astronomy, particularly where new frontiers are being opened up? Does this sug- gest that there are important aspects to physical theory which we still understand so poorly because of classical prejudices fostered by the twentieth century meta-theoretical tradition of philosophy of science?

Since the answer to this question is in the affirmative, we would like to consider how and why the opposition to the methodology of inquiry resulted in its *exclusion* from this field of research. We are mainly interested in exploring how it might be possible to bring it *back in.* Thus our discussion will cover research in physics and in the philosophy of physics, how they have been connected in the twentieth century and how they should be connected or reconnected in the twenty-first century. For our purpose, where necessary, we may use the terms "context of discovery" and "context of inquiry" synonymously. But first of all, let us find out what is conventionally or traditionally meant by the context of discovery. And let us consider the implications of the claim that the methodology of inquiry (in the context of discovery) is one of those things with which a philosopher of science can have no concern at all.

There may be some merit in thinking of it - in the meta-theoretical tradition of the twentieth century analytic philosophy of science - as that which is *external* to a science such as physics. But this already depends on our decision as to what is it in physics which the philosopher of physics can or should study. Since Kant, physics is taken by philosophers in the context of its concrete achievements, its finished theories and their appraisal or evaluation in terms of their performance when confronted with experimental evidence. Everything else falls *outside* the preferred context of justification, taken as the context of the scientific rationality of theory-choice. Thus, strangely enough, not only the cultural matrices which are presupposed by the disciplinary matrices of physics, but also the physicist's search for the kind of theory which physics can aim at, given a certain epoch of its development, are excluded from the domain of philosophical reflection and analysis. This only indicates how artificial and dogmatic our philosophical conception of physics, and of philosophy of physics, is bound to be, if we do not carefully examine the assumptions that lie at the very roots of our thinking about science. The philosophical game of *inclusion and exclusion* runs its own risks, particularly the risk of gravely distorting from the very outset the proper subject of philosophical investigation. We want here to argue that this is the point where the twentieth century philosophy of science has failed. In restricting its domain of investigation to the context of theory-justification, or theory-choice, in the sciences, and in *excluding* the context of discovery, it has preferred to remain ignorant of the most fascinating aspect of scientific rationality. It is this very aspect of scientific rationality which we propose (sect. 2.3.) to designate as the dynamic core-context-building in physics.

Coming back to the main subject of our discussion, we can explain the opposition to the methodology of inquiry in two ways: (i) First, by showing what misunderstandings about science, such as physics, underlie this opposition; and (ii) secondly, by deriving it as a simple consequence of a certain kind of approach to connecting science to the philosophy of science.

We have mentioned the great divide between scientific theory itself and philosophy of science conceived as the meta-theory. The great divide demanded that philosophers and methodologists should articulate scientific theories only when considered *in the context of justification*. This implied that, beyond this, philosophers could not legitimately ask questions concerning scientific theories if taken in the *context of development or discovery*. This restriction clearly *excluded* any concern with *theory-development* in physics. It was also dogmatically assumed that, if taken in the context of discovery, scientific theories would be neither available nor accessible to proper logical analysis. Interestingly enough, this turned the availability of complete theories into a necessary condition for *rationally reconstructing* the rationality of science from the perspective of its explanatory goals. *Where complete and readily available*, the scientific theories were to be subjected to *logical analysis* in terms of pure logical forms and logical connections shared by their statements and by the reports of experiments. Notice that the rational reconstructions addressed the scientists themselves as following well-defined rules in accepting or rejecting proposed theories, while taking well-articulated experimental evidence into account in doing so. Restricted to the *static* categories of logical analysis, the great divide was in a sense a re-incarnation of the older doctrine of the subject-object dualism. However, in the meta-theoretic tradition, the logic of science was intended to be ontologically neutral. There is no doubt that this tradition still flourishes and even may dominate the philosophy of science at the beginning of the twenty-first century.

We can now show the consequences this has had for philosophical research within the science studies. In particular the following serious consequences call for explanation: (i) That no attention has been paid to the physicist's search for the kind of theory which physics can aim at, whether in the context of the methodology of unification of different laws of interaction, which physics has consistently employed, or in the context of the methodology of seeking better alternatives to extant theories. (ii) That no research on the dynamic core-contexts of development, or on the theory-experiment interface, has been possible as a result of the exclusion of the context of discovery from research. (iii) That, on a rather wider scale, no significant research has been done recently on the dynamic relationship between the disciplinary matrices of a science such as physics and its cultural matrices. (iv) That the dogmatic, rigid, and artificial nature of the traditional distinction between the context of discovery conceived as operating from outside a science *and* the context of justification has not been itself exposed to any decisive criticism. And (v) since rational analysis is only applied to static aspects, the progressive succession of theories was either considered to be irrational (e.g., by Kuhn) or rather fortuitous (as by Popper).

There is another serious consequence that deserves attention. "Most twentieth century philosophers of science," laments Nicholas Maxwell (2006: 1), "assume without question that they pursue a meta-discipline - one that takes science as its subject matter, and seeks to acquire knowledge and understanding about science without in any way affecting, or contributing to, science itself. (This continues to be the case in the first years of the 21st century)..." As regards his expression of disappointment, Maxwell is not alone. The question raised by him whether philosophy of science should contribute to science directly deserves serious attention. But it must be recognized that it is a matter of debate how high our expectations can be regarding philosophy of science as to its potential for impacting or "contributing" to science. If we only think of the dominant philosophical accounts - whether in the Carnapian, Popperian, Kuhnian or Lakatosian tradition - of scientific rationality which take no interest in the active frontiers of physical theory, particularly where these interface strongly with the core-contexts of theory-development, it is no surprise that philosophy of science is of no direct help to science. It cannot be expected to make direct contributions to science.

2.2.1 Weinberg's Critique of the "Unreasonable Ineffectiveness of Philosophy"

In the kind of situation in which philosophy of science finds itself throughout the twentieth century, it may not be very surprising, if we find that the eminent physicist Steven Weinberg, (Nobel Laureate, physics 1979) has devoted a whole chapter of his book *The Dreams of a Final Theory* (1992, 166-190) to a discussion on physics and philosophy, contrasting "the unreasonable effectiveness of mathematics" (Wigner 1960:1-14) with "the unreasonable ineffectiveness of philosophy" in the development of science. Weinberg's lament is largely appropriate for the twentieth century philosophy of science. His negative attitude against the latter becomes already clear from the very title of one of the chapters: "Against Philosopphy". Though we think that many of the conclusions drawn by Weinberg are not justified, it is certainly worthwhile to analyze the reasons for his negative attitude and to explore how it might be possible to meet his challenge.

Weinberg (1992: 166) poses the following question for physicists to ponder over: Can philosophy give us any guidance toward a final theory? His own answer is as follows: "It is just that knowledge of philosophy does not seem to be of use to physicists - always with the exception that the work of some philosophers helps us to avoid the errors of other philosophers".

Like science, philosophy cannot afford to turn a blind eye to the question how much accessible the works of philosphers are to non-philosophers, particularly to scientists. In this very context, Weinberg (1992: 133-134) observes:

> "From time to time ... I have tried to read current work on the philosophy of science. Some of it I found to be written in a jargon so impenetrable that I can only think that it aimed at impressing those who confound obscurity with profundity. ... Only rarely did it seem to me to have anything to do with the work of science as I knew it ... I am not alone in this; I know of no one who has participated actively in the advance of physics in the post-war period whose research has been significantly helped by the work philosophers".

He recalls that during the first years of his studies at Cornell University he was eagerly following courses of philosophy and that his anti-philosophical affection was born in him by an over-saturation of philosophy. He does not write which courses of philosophy he was following, but it is very likely that it was philosophy of science. It is interesting that around the same time one of the great physicists of the twentieth century, R. Feynman, was Professor of physics at Cornell and he too wrote very bluntly on his colleagues in the philosophy department, saying that "the guys from this department were particularly inane" (Feynman 1985, p. 232).

Weinberg admits that it is not absurd to believe, from the very beginning, that physics, in which rather vague criteria like "beauty of a theory" play sometimes an important role, could receive some help from philosophy. But, with the overwhelming majority of physicists, he concludes that it is no longer the case. As this has also been acknowledged by many philosophers, he quotes Wittgenstein as saying that: "nothing seems to me less likely than a scientist or mathematician who reads me should be influenced in the way he works". This could be the cause of a neutral attitude to philosophy in general, and of hostile attitude only to those schools which try to teach physicists how they should proceed (e.g., as some of the constructivists do).

Weinberg acknowledges that in the earlier periods physics did draw some advantage from philosophy. It is generally recognized that the positivism of Ernst Mach had an important effect in physics at the turn of the nineteenth to the twentieth century. Taking up the scepticism of Hume,[3] Mach's main concern was to question the uncritical attitude towards the mechanistic world-view and the concepts of space and time of Newtonian physics. This was a decisive step on the road to the theory of relativity, as is acknowledged by Einstein (1916, p. 101) in his Epilogue on Mach. Positivism was more popular among physicists from the continent and especially central Europe than among those in Great Britain and this may have been the reason that it took some time until the special relativity was acknowledged there. An anecdote is very revealing and instructive in this respect: Around 1910 Ernest Rutherford, a typical exponent of Anglosaxon pragmatism, was twitting Willy Wien about relativity. Wien said then to Rutherford that no Anglo-Saxon could understand it. No, Rutherford replied laughingly, they have too much common sense (Pais1986: 190).

Perhaps even more important was the influence of positivism on the development of one of the greatest achievements of twentieth century physics, quantum mechanics. This is fully acknowledged even by Weinberg. He writes: "Positivism played also an important part in the birth of modern quantum mechanics... In the spirit of positivism, Heisenberg admitted into his version of quantum mechanics only observables ... The uncertainty principle ... is based on Heisenberg's positivistic analysis of the limitations we encounter when we set out to observe a particle's position and momentum".

After these important contributions of positivism, the harm done by it seems rather small. An argument brought forward quite frequently is Mach's opposition against atomism. It is true that Mach was fighting against the mechanical atomism at the end of the 19th century. But his was only a minority view. After all, the picture of the atom as it turned out after establishment of quantum mechanics is quite far away from the mechanistic atoms in vogue before.[4]

Given the admitted benefits of philosophy, the critique of Weinberg against philosophy seems exaggerated. But before coming to a possible explanation, let us contrast this anti-philosophical view with the opinion of the scientists who themselves can be regarded as philosopher-scientists.

In his speech for the benefit of a memorial for Immanuel Kant, H.

---

[3]The principal oeuvre of Ernst Mach: "Mechanics, Represented Historical-Critically" is dedicated to the memory of David Hume.

[4]It is true that also Max Planck criticized E. Mach severely. Presumably, this was so because this critique prevented him for a long time to adopt atomism (R. Joos, private communication).

Helmholtz, who saw himself attacked for allegedly anti-philosophical attitudes[5], made a famous statement that "to investigate the sources of our knowledge and the degree of its justification is a duty, which will ever be that of philosophy and no century can avoid that unpunishedly".[6] Helmholtz was living up to this duty and his classical books on physiological optics and acoustics have frequently decisive passages just devoted to this duty. Also H. Poincare gave epistemological reflections a very high value, as can be seen from the fact that he was publishing three important books for that purpose.[7] Here we have mentioned only the giants in the field, but epistemological reflections were common in the work of many of the important scientists, at least of those from the continent. Some fifty years ago a remarkable reminder of all this came from Henry Margenau (1949) in these words: "Every discoverer of a new physical principle makes an important contribution to philosophy, even though he may not discuss it in philosophic terms".

Let us now try to compare the seemingly contradictory opinions on philosophy. As mentioned above, Weinberg recognizes that there might have been some positive influence of philosophy on science. But he claims that this was mainly due to the fact that it could shield physics from other harmful philosophical prejudices. In some sense one may say that this is similar to Helmholtz's demand for an enquiry into the sources of our knowledge and the prevention of dogmatic prejudices, as exemplified in Kant's *Critique of Pure Reason* (1781, 1787). One should be aware, however, that even the "common sense", evoked in the discussion between Wien and Rutherford mentioned above, is a dogmatic philosophical prejudice and has hindered progress in science at many instances.

In the sense that philosophy of science should never be dogmatic but only critical, one may say that there is no contradiction between Weinberg and the great scientist-philosophers of the 19th century.

Another important clue to understanding Weinberg's criticism is his remark that philosophical doctrines, even if helpful for a certain time, live in general too long and do at the end more harm than good. In this respect it is especially revealing that a main target of his attacks is

---

[5]He was indeed, as most good scientists of the time critizizing heavily the "Naturphilosophie" of Schelling and Hegel.

[6]H. Helmholtz (1884, Bd I, p.368).

[7]H. Poincare (1902, 1905, 1909).

positivism and its off-springs. He writes that his main targets in the chapter "Against Philosophy" were positivism and conventionalism, but that the title "Against Positivism and Conventionalism"[8] would not have been particularly catchy. We concentrate on positivism and, therefore, in order to deal with the reservations of Weinberg against philosophy we have especially to reconsider the offsprings of Machian positivism, e.g., Carnap's philosophy of science on which we shall comment later in Chapter 5.

It is true of many philosophical accounts of science that have been produced in the twentieth century, with an approach to science usually chosen from *outside* science, that they do not explain its most fascinating aspect, namely its success if averaged over periods of say half a century. If we follow the great examples of Pierre Duhem and Ernst Cassirer, asking how close and near they stay to physics, we will find that one of the most important tasks for the methodology of science is to understand theory change by theory development in physics, a task which it can fulfil by paying attention to the deeds of scientists *inside* physics rather than their statements *about* physics. In this context, it is particularly challenging to understand theory-development in terms of *the continuity of the formalism* across the possible radical conceptual changes or rather scientific revolutions. For what this suggests is that the development of science, i.e., theory-development, is made

---

[8]Here it is rather the sociological conventionalism which is meant and not the epistemological conventionalism of Poincare. If there is a trivial truth about science it is the truth that science is a social activity, or an activity which is rational in so far as the scientists employ the appropriate means to achieve the appropriate ends. However, with Steven Weinberg, we want to say that the methodology, or epistemology, of science as a normative enterprise has nothing to do with the sociology, or even psychology, of science. While rejecting the idea that 'our theories are not much more than social constructions, as supposed by some radical commentators on science, such as Pickering, the author of a book entitled *Constructing Quarks*', Steven Weinberg (Weinberg 1997, 42) says that, 'We know of course that science is a social activity. As Latour and Woolgar commented after observing research in biotechnology, "The negotiations as to what counts as a proof or what constitutes a good assay are no more or less disorderly than any arguments between lawyers and politicians". But the same could be said about mountain climbing. Mountain climbers, like biochemists or lawyers, may argue over the best path to the peak, and of course these arguments will be influenced by the traditions of mountain climbing and the history and social structure of the expedition. But in the end the climbers will either get to the peak or they will not, and if they do get there they will know it. No mountaineer would write a book about mountain climbing with a title like "constructing Everest" '.

possible by core-context-building and re-building from *within* science rather than by radical conceptual changes or by the intervention of different kinds of contexts of development from *outside* science (see sect. 2.1.3 and Chapter 4, where the concept of dynamic core-context of development is introduced and discussed).

2.2.2 Bridging the Divide

It is important to distinguish between the two kinds of *contexts* of scientific development. On the one hand, corresponding to the context of discovery in the conventional sense of the philosophies of science of logical empiricism (Carl Hempel, Rudolf Carnap, Hans Reichenbach, among others) and critical rationalism (Karl R. Popper and his followers),[9] there are contexts of development which are *external* to physics. Sometimes, they are also referred to in the discussions on *science in context*. Contexts of development that are *external* to science, e.g. to physics and astronomy, do not play a direct role in the methodology of theory-development and unification.

On the other hand, there are contexts of development which are internal to physics. They belong to its dynamic core-contexts of theory development, as we have designated them in this book. Philosophers of science may have reluctantly recognized the former, i.e., the external contexts, and yet chosen to exclude them from philosophical study. Their exclusion from the rationality of science, which is taken as the subject of philosophical reconstruction, in *turn* prevented the most leading twentieth century philosophers of science from recognizing and studying other important aspects of physical theory, e.g. the contexts of development that are internal to science. It is these contexts of development that are internal to physics and astronomy which play a dynamical role in the methodology of physics. But these have not received the attention which they deserve.

Unlike the analytic philosophers of science of the 20th century, we do not think that it is possible to exclude even "science in context" from serious philosophical study. What is not clear from the philosophical accounts of scientific development is this: That at any given time there may not be a single factor but a whole diversity of con-

---

[9]They all required philosophy of science in the analytic tradition to exclusively focus on the *contexts of justification*, also called contexts of theory-appraisal/theory-choice, in order to reveal the logical structure of science which plays a crucial role in scientific explanation.

textual factors which influence scientific development from *outside*. Consider, e.g., factors like state's science policy, the problem-solving skills of the scientific community and the funding of scientific research. It has been the normal practice among philosophers of science to give them the collective name of "context of discovery", which is too over-simplifying and misleading for any legitimate debate on science in *context*. Although philosophers prefer to study the logical or episte-mological or methodological aspects of science, it remains an important philosophical task to articulate the diverse contexts which impact scientific development.

In the 21st century, philosophy of science should take up the task of addressing hitherto unforeseen new challenges, including the challenge of articulating the external contexts of scientific development which impact every science deeply. In this sense, philosophy of science can make decisive contributions to public understanding of physics-in-context. Think of the most important yet highly problematic areas like science policy, the state beaurocracy and its complicated mechanisms of control over science and scientific establishments, private and public investment in research and development, innovation in applied science and technology, and (the rarely explored) possibilities of investment in research on *access* to new technologies (see Pandit 2008a). Any debate on the issues relating to these areas would enrich public understanding of how a diversity of factors impact physics and other sciences *from outside*. There can even be a high expectation from philosophy of science in these areas, more so where questions of ethics and governance are concerned.

The question may be asked: How high should our expectations really be? Philosophy of science has first of all to cope with the doctrine that there is a great divide between it and science, a doctrine which the twentieth century philosophers of science working within the analytic tradition have commonly fostered. This may be taken to imply that "science in context" is beyond its scope. If we give up this doctrine along with this implication, we might change the whole question beyond recognition. Consider the following possibility. We might quite reasonably like to think of the methodology of physics as being continuous with core-context building and re-building in physics. After all if there is a fundamental aspect of physics which should be of great interest to philosophers of science, it is the methodology of physics understood as a theory of scientific change and theory development.

Understood in this sense, it would make no sense to assume, first, that the methodology of physics is an external discipline and then to ask the following question: How does it contribute to physics? That is to say, on our rejection of the above doctrine, one cannot think of dynamic core-context building in physics without thinking of the methodology of physics. Therefore, there is no question how the latter contributes to the former, the two being so dynamically related with each other.[10]

Conceived within a different philosophical framework altogether as the present study is, it departs from the twentieth century analytical tradition in more than one way. This entails a change in approach at various levels of analysis as follows. *First*, viewed from a *dynamical* instead of a static perspective, scientific theories cannot be articulated in terms of the old dichotomy of the *context of justification and context of discovery*. No doubt, this dichotomy has proved very efficient in producing *forms of exclusion* within the science studies. For example, it excludes theory-development from serious study altogether. The great divide between *the scientific theory* and *the meta-theory* sustained by this dichotomy breaks down completely if theory-development is incorporated within the methodology of physics. *Secondly*, instead of dogmatically accepting *the great divide*, we must recognize that there is *continuity* between physics and the methodology of physics, full of feedback-influences. The *continuity* provides a major resource for connecting and re-connecting the one with the other. *Thirdly*, philosophy (and methodology) of physics should constantly guard itself against the pre-dominant tendency among philosophers of science to resort to the strategy of changing the subject of scientific appraisal

---

[10]As a student, having studied Robert Sanderson's *Logicae Artis Compendium* (Oxoniae, 1618) along with various works of Rene Descartes, each dealing with method, differently, Newton himself distinguished between the method of invention and the method of doctrine, or the method of discovering knowledge and the method of presenting and teaching it. In the light of this distinction, we can say that philosophy of science cannot be expected to be *knowledge-producing* in the standard sense in which physics is knowledge-producing (Pandit 2007). Therefore, it cannot be expected to contribute to the core-context-building and re-building in physics. At the most, it can indirectly help in advancing this very fundamental aspect of physics by interrogating the method of presenting and teaching scientific discoveries and by interrogating scientific progress from one core-context to the other. Following Einstein's advice (1934, 113), we might express this as follows: Philosophy of science can best contribute to our understanding of physics by paying critical attention to the deeds of (= the work produced by) the physicists and not just to their words.

in science, viz., scientific theory itself, in order to achieve the results that are desired from the very beginning (Pandit 1982, 1991). Thus, while there is no reason to doubt their serious commitment to studying and analysing *the structure of physical theory*, they invariably land themselves elsewhere. We find them discussing subjects not even remotely connected with the subject of scientific appraisal (Pandit 2008b). Moreover, they develop criticisms of other approaches in terms of the kind of categories which no one outside *the group* can understand. *Fourthly*, within physics, *continuity* reigns supreme. Until now, continuity in the development of physical theories implied that a new theory should, under certain conditions, reproduce the successes of the theory which is being superseded. It shows itself most prominently in physics, as its later theories replace its earlier theories. At a deeper level, however, theory development and unification are informed by *core-contextual continuity* (turn to discussion in Chapters 4 and 5). *Continuity* in this sense is characteristic of *the dynamical complexity of physical theory*. It shows itself more profoundly at the theory-experiment interfaces at the active frontiers of development in physics. *Fifthly*, it is well-recognized since the time of Newton that physics progresses by its movement *backwards* from problems to theories, replacing the theories of a smaller range of applicability by those with greater range.

Sixthly, besides theory-succession in the standard sense just alluded to, there is need for researching how physics progresses even more effectively by its movement *forwards* from theories to problems, opening up new frontiers (Pandit 1982, 1991). It is here that *the core-context building resolving power of physical theories* comes into play in a fundamental way. Taken in this fundamental aspect, physical theory clearly falls outside the scope of a purely logical analysis of science, focusing on its syntactical or semantical aspects. *Seventhly*, at a deeper level, the possibility of wide-ranging progress and growth of knowledge within physics has to be understood in terms of the methodology of dynamic *core-contexts of theory development and unification* which is the subject of discussion in Chapter 4. *Lastly*, the model of theory development and theory unification which we propose to formulate in this study can be extended to technology in general and to the theory-experiment interface in particular. In physics, this consequence is particularly important in the context of the role of instrumentation, experimental design and construction of particle accelerators and detectors across the world .

Physics-in-context which we are here referring to may, in a sense, seem to be akin to Thomas Kuhn's (1962, 1974, 1977a) paradigms-driven or disciplinary matrices-driven normal science. In Kuhn's picture, on the one hand, it is these paradigms, understood as exemplars, which confer power, authority and legitimacy on the community of scientists who are supposed to practice the puzzle-solving normal science. But, on the other hand, it is the scientific community as a whole, its activities, tools, commitments and tacit rules of understanding which form the paradigm in an all-embracing metaphysical or sociological sense. This calls for a detailed look at Kuhn's picture of science. In the following section, we shall not only briefly indicate how his views differ from our own but demonstrate that his approach provides yet another illustration how science might be studied from a perspective which lies quite outside science (for more details turn to Chapter 6).

2.2.3 Scientific Change: How Not To Understand Incommensurability

Since meaning-variance in science is part of theory-change, theory-development, and theory-succession, the very idea of "incommensurable" theories, i.e., theories without successors, as it dominates the work of Kuhn and his followers, is totally inappropriate for science and science-studies. Yet it is this very idea which has shaped the discussions of the nature of theory-change in Kuhn's picture of science as it has evolved from relatively simpler to complex versions over a long period of time (Kuhn 1962, 1970a, 1970b, 1974, 1977, 1981, 1983, 1989a, 1989b, 1990a, 1990b, 1991a, 1991b, 1992, 1993). While coping with his critics and admirers alike, Kuhn kept on groping for newer metaphors where his older metaphors had outlived their utility (see Franklin 1999, 276-281). Exchanging metaphor after metaphor, as in his "Second Thoughts on Paradigms" (1974) followed by his "The Road Since Structure" (1991a, pp. 3-13) and "Afterwords" (1993), Kuhn first moved from paradigms-driven puzzle-solving science to disciplinary-matrices-driven science and then from language, communicability and translatability to lexical taxonomy (or kind-terms) and localized incommensurability due to localized meaning variance (Kuhn 1983). Thus, at one stage Kuhn (1976, p. 191) reformulated the problem of "incommensurability" as follows:

In applying the term 'incommensurability' to theories, I had intended only to insist that there was no common language within which both could be fully expressed and which could therefore be used in a point-by-point comparison between them.

We must not miss here Kuhn's emphasis on language of science as a conceptual framework. Moreover, in Kuhn's sociological account there are inherent many complexities, notably in his doctrine that a paradigm is "not simply a particular scientific theory but a whole way of working, thinking, communicating, and perceiving with mind. It is based largely on the skills and ideas that are tacitly transmitted during what could be called a scientist's apprenticeship, in graduate school for example" (see Bohm et al. 1987, p. 52). Because of the enormous complexity which is built into the notion of scientific community, its activities, tools, aims, commitments, its belief systems, its tacit rules of understanding, its institutions and practices of training young minds, and its behavioural standards, such an all-embracing sociological approach to paradigms and their alleged role in science does not allow a paradigm to be defined as precisely as one would wish. This aspect of Kuhnian paradigms is further accentuated by "the main force of Kuhn's idea, which is that the tacit infrastructure, mostly unconsciously, pervades the whole work and thought of a community of scientists" (see Bohm et al 1987, p. 52). Thus, if we follow Kuhn strictly, we can neither identify a paradigm with a fundamental general theory nor identify a change of paradigm with a consciously produced change in this theory. As a consequence, neither theory-development nor growth of scientific knowledge can be articulated properly in terms of paradigms.

In recent years, some philosophers, notably Nersessian (2001, pp. 275-301), Shapere (2001, pp. 197-198) and Pandit (1982, 1991),[11] have argued to the effect that "incommensurability is not a problem in and of itself".

Similarly, commenting on how Kuhn has gradually moved away from initial statements implying incommensurability between

---

[11]Arguments of Nersessian (2001) and Shapere (2001) are somewhat similar to the train of arguments in (Pandit 1982, 1991), which showed, for the first time, how Kuhn has resorted to the strategy of changing the subject of scientific rationality, making the scientific community rather than physical theory the subject of scientific appraisal and creating the non-issues of theory-incommensurability.

paradigms, while increasingly seeing the problem of incommensurability as more "local", Shapere (2001, p. 197-198) argues: "But in all this evolution, there has been surprisingly little attention to a possibility that seems quite appealing: namely, that two uses of the same term in two distinct theories (for example, in an earlier and a later theory) may be quite different, and yet be related by a chain of reasoning which explains why the term's meaning (and/or reference) had to be changed. The source of this neglect is perhaps evident in the views from which the problem of comparison stems: for if what is counted as a reason in a particular scientific community or tradition is dependent on a "paradigm", or on more specific and distinct exemplary models of problem-solving, the concept of a reason is derivative, and perhaps even varies (perhaps in a radically incomparable way) from one community or tradition to another. I think this has been a mistake: the important task for philosophy of science is not to make it a branch of linguistic philosophy, but to examine what is certainly the most important claim made by science, that its methods and conclusions are based on reasons.

If linguistic analysis is thus committed to the flames as the fundamental way of understanding what happens in inquiry, and if scientists come to their conclusions in case of major change of theory and/or theoretical perspective by employing relevant prior knowledge as reasons in particular inquiries, then the problems of incommensurability, like so many others, disappear. There will still remain differences, often radical, between two theories and their uses of similar or identical expressions. But is that not what we expect of science, which continually revises not only its meanings and references, but also its conception of what it studies (its domains), the kinds of explanatory accounts to give of what it studies, its methods, standards, goals, and so much else?"

On the contrary, the problem of theory-incommensurability arises "from approaching problems pertaining to science as though they were problems pertaining to languages alone"(Nersessian 2001 pp. 275, 298). In other words, it arises by

> "placing too much analytical focus on scientific conceptual structure as languages and so transferring to science what might be said of languages generally" (Nersessian 2001, p. 277),

Here, we want to point out that the only possible rationale why Kuhn chooses to view science and scientific change through the lens of "incommensurability" may be sought in his approach to science studies from outside science, viewing it as if it was just a language among other languages or a language community among other such communities (Kuhn 1976, p. 191). No doubt, Kuhn's view of science demands a closer look than is possible in the present discussion (for more details turn to Chapter 6). However, in order to indicate possible dissimilarity or similarity between our approach and Kuhn's approach, it is important to mention those points which emerge from Kuhn's (1991a, pp.3-13) changing views, some of which may even seem at first promising.[12]

First, in a sense Kuhn (1991a, pp.3-13) is right in saying that the assumption, often made in the discussions of rationality and relativism, that what is at stake in such discussions is "the correspondence theory of truth, the notion that the goal, when evaluating scientific laws and theories, is to determine whether or not they correspond to an external, mind-independent world" must "vanish together with foundationalism". And secondly, we find Kuhn's (1991a, pp. 3-13) following new point regarding what happens after a revolutionary scientific development only partly in harmony with our approach (see Pandit 1982, 1991) but partly unacceptable where Kuhn draws far-reaching and strong parallels between biological evolution and evolution of scientific knowledge:

> After a revolution there are usually (perhaps always) more cognitive specialities or fields of knowledge than there were before. Either a new branch has split off from the parent trunk, as scientific specialities have repeatedly split off in the past from philosophy and medicine. Or else a new speciality has been born at an area of apparent overlap between

---

[12]Here we shall not discuss Kuhn's following claim that an evolutionary epistemology need not be a naturalized one, being highly debatable (Pandit 1982, 1991): "When I first got involved, a generation ago, with the enterprise now often called historical philosophy of science, I and most of my coworkers thought history functioned as a source of empirical evidence. That evidence we found in historical case studies, which forced us to pay close attention to science as it really was. Now I think we overemphasized the empirical aspect of our enterprise (an evolutionary epistemology need not be a naturalized one). What has for me emerged as essential is not so much the details of historical case studies as the perspective or ideology that attention to historical cases brings with it" (1991a, pp. 3-13).

two pre-existing specialities, as occurred, for example, in the cases of physical chemistry and molecular biology. . As time goes on, one notices that the new shoot seldom or never gets assimilated to either of its parents. Instead, it becomes one more separate speciality, gradually acquiring its own new specialist's journals, a new professional society, and often also new university chairs, laboratories, and even departments. Over time a diagram of the evolution of scientific fields, specialities, and sub-specialities comes to look strikingly like a layman's diagram for a biological evolutionary tree. Each of these fields has a distinct lexicon, though the differences are local, occurring only here and there. There is no lingua franca capable of expressing, in its entirety, the content of them all or even of any pair...

Now coming to the parallels Kuhn (1991a, pp. 3-13) draws between biological evolution and evolution of knowledge, consider his following statement:

First, revolutions, which produce new divisions between fields in scientific development, are much like episodes of speciation in biological evolution. The biological parallel to revolutionary change is not mutation, as I thought for many years, but speciation. And the problems presented by speciation (e.g., the difficulty in identifying an episode of speciation until some time after it has occurred, and the impossibility, even then, of dating the time of its occurrence) are very similar to those presented by revolutionary change and by the emergence and individuation of new scientific specialities.

But we strongly disagree (see the Fig. 2.1 in sect. 2.2.5., contrasting Kuhn's tree of evolution with our methodological model of the core-context of development) with Kuhn (1991a, pp. 3-13) as regards the second parallel which he (1991a, pp. 3-13) draws between biological and scientific development, which concerns the unit which undergoes speciation (as distinct from a unit of selection). It is more striking yet reminiscent of his earlier emphasis on the puzzle-solving scientific communities. Consider the following statement:

In the biological case, it is a reproductively isolated population, a unit whose members collectively embody the gene pool which ensures both the populations' self-perpetuation and its continuing isolation. In the scientific case, the unit is a community of intercommunicating specialists, a unit whose members share a lexicon that provides the basis for both the conduct and the evaluation of their research and which simultaneously, by barring full communication with those outside the group, maintains their isolation from practitioners of other specialities.

A brief comment is in order here. In the linguistic imagery of science drawn by Kuhn, he seems to be so over-whelmed by the jargon (see sect. 2.2.4 below) that he completely misses the role of mathematics, and more importantly the continuity of formalism, as a universal "language of science". He also misses the whole point of how divergent fields of science, with different languages, converge and come together. Consider, e.g., the coming together of particle physics and cosmology in recent decades. Before that, quantum mechanics and chemistry have already set an example. On the contrary, what is allowed to dominate in Kuhn's imagery is the image of a historically given language community whose members are able to interact and participate in its cultural (even "disciplinary matrix-driven") life-world. Simultaneously, the language, in this latter sense, distinguishes, separates and isolates one community from other language communities (see Pandit1991, pp. 55-80). The question is whether all that which is true of language generally can be transferred to the scientific groups as is being done by Kuhn (1991a, pp. 3-13):

> "It is groups and group-practices that constitute worlds (and are constituted by them). And the practice-in-the-world of some of those groups is science. The primary unit through which the sciences develop is thus the group, and groups do not have minds".

At this point, a re-incarnation of Kuhn's (1962, 1970b) doctrine of the primacy of the puzzle-solving scientific community over its members appears in Kuhn (1991a, pp. 3-13):

> The primacy of the community over its members is reflected also in the theory of the lexicon, the unit which embodies the shared conceptual or taxonomic structure that holds the community together and simultaneously isolates it from

other groups. Conceive the lexicon as a module within the head of an individual group member. It can then be shown that what characterizes members of the group is possession not of identical lexicons, but of mutually congruent ones, of lexicons with the same structure.

If incommensurability is taken in a generalized sense, we must remember that one has to take into account the fact that it is a relative notion, since it depends on which common measure is chosen. This becomes clear if one sticks to its original definition, viz.,: Two numbers are said be incommensurable if their ratio is not a rational number. The discovery that the diagonals of a square and its sides are incommensurable provoked a deep crisis in Pythagorean mathematics, since the latter was based on whole numbers. If the notion of numbers is however generalized in an intuitive way, namely into real numbers, there is no longer an essential difference between the irrational and rational numbers, and incommensurability, in the generalized sense, just disappears. So "incommensurability" is clearly a relative notion. Let us consider absolute time in Newtonian mechanics and relative time in special relativity as an example. First, if viewed from outside, both seem incommensurable. If viewed within the scheme of special relativity, Newtonian mechanics, in which "absolute time" enters as a parameter, can be seen as a special case of special relativistic mechanics. Therefore, in retrospect, the precursor theory is in a significant sense contained in the successor theory. Secondly, and what is more important, the operational definition of time, so to speak the time in the context of physics, was less affected than the philosophical" one. Though Newton uses strong words in introducing the mathematical or true time as distinct from time as an everyday notion, if it comes to substance, he brings in as important example the fact that the sideral time is more suited for scientific considerations than the solar time, since the former is closer to the mathematical time. This is certainly true both in relativistic and non-relativistic mechanics. In semiotic terminology (Dosch et. al 2006), incommensurability applies only to the connotational, but not to the proper scientific or denotational, level.

Having said this, we want however to stress another point as follows: The transition between the so-called "incommensurable" theories plays an important role in the dynamic core-context of scientific development, since often they create new problems, opening up new frontiers, which are solved only much later. Coming back to our original example above, namely the incommensurability of the diagonals and the sides of a square, the problem created by it in ancient times was treated in a fully adequate way only in the 19th century and the theory of real numbers was brought into its final form by Cantor (N. Bourbaki 1984). Very often, the problems created by the transition between "incommensurable" theories stay throughout the history of science, repeatedly giving rise to new problems and building interfaces. The problem of locality is a good example. Locality played an important role in modern science. Many continental scientists did not accept Newton's theory of gravitation, since it seemingly involved an action at-a-distance. Newton himself never claimed that it was such an action. On the contrary, he considered it as

> "inconceivable that inanimate brute matter should, without the mediation of something else ... operate upon and affect other matter without mutual contact".

Nevertheless, slowly the concept of action at-a-distance was accepted and it took quite some time until the concept of strict locality gained again general recognition. In the axiomatic formulation of relativistic quantum field theory, the principle of locality plays a fundamental role (see e.g., Streater and Wightman 1964). Now it has again come in the focus of fundamental research with the rise of inherently non-local superstring theory as a candidate for a theory superseding local quantum field theory.

In our view, in order to understand the diachronic nature of theory-development in the sciences such as physics, we can dispense with those models of scientific progress, whether empiricist or rationalist, that share the following characteristics in common:

(i) They assume that it is always possible to choose, from outside it, an aim for science (e.g., Popper 1983, xix-xxx; van Fraassen 1980, 1991);

(ii) they further assume that in order to articulate scientific rationality it is necessary to deal with scientific theories that are complete with their axiomatic foundations, implying that it is not possible to deal with theory-development;

(iii) they require that, once an aim for science is chosen from outside, of any two or more rival theories in their field the one which achieves the aim of science better than others can be the best candidate for rational choice; and/or

(iv) they require, using criteria from outside science, that conceptual, or connotational, discontinuities, meaning-variance and theory-incommensurability matter a lot in understanding how scientific change occurs.

We can dispense with such models in so far as we want to keep close to the inner dynamics of core-context guided theory development and theory-unification in science such as physics (Pandit 1982, 1991, 2002a, 2002b, 2007b). For example, it is not possible to recognize the importance of the gauge principle in particle physics without resorting to the methodological role the dynamic core-context plays within physics. It is one thing if the philosophers of science choose an aim for science from outside, as Popper (Popper 1983, xix-xxx) and Bas van Fraassen (1980, 1991) do. But it is a different matter altogether to recognize how the core-context guided theory-development and theory-unification are quite characteristic of the sciences of physics and astronomy. For this reason we are inclined to agree, if to a limited extent only, with the following point made by Kuhn (1991a, pp. 3-13) which he traces to the parallel between scientific and biological development he suggested at the end of the fist edition of *The Structure of Scientific Revolutions* (1962) without an explicit diachronic perspective:

> scientific development must be seen as a process driven from behind, not pulled from ahead - as evolution from, rather than evolution towards. In making that suggestion, as elsewhere in the book, the parallel I had in mind was diachronic, involving the relation between older and more recent scientific beliefs about the same or overlapping ranges of natural phenomena.

2.2.4 The Language of the Scientific Community

The expression "scientific language" is too simple to describe the full linguistic and symbolic complexity of the language of a scientific community. This complexity can be articulated by drawing a three-fold distinction between three kinds of component languages (see Dosch 2011): Everyday language, the scientific language and the "scientific jargon". The "scientific language" is well-defined and codified, and it is common to all scientific communities. A "Fourier transformation", e.g., is understood by an electrical engineer, a particle physicist and a neuro-physiologist in the same way (albeit with a difference in emphasis on mathematical rigour). "Scientific jargon" refers to "scientific language" where it is diluted by everyday language. Thus, both "scientific jargon" and "scientific language" are embedded in everyday language.

Unlike "scientific language", "scientific jargon" is very specific to a specialized scientific community. It is partly the scientist's linguistic indifference which leads to it, but, much more important, it contains pictures which can be derived in a more or less rigorous manner from formalism. Normally, these pictures are very useful in some applications, while they fail completely in others. For example, the statement "elementary particles consist of quarks and gluons" belongs, strictly speaking, to scientific jargon, since the formalism behind this statement is very complex. Statements like "in this process a virtual pion is created" also belong to scientific jargon, however well-embedded they may be in the formalism of quantum field theory. It would be a typical example of what Whitehead calls "misplaced concreteness" to make such a statement the starting point of a philosophical discussion on the "existence" of virtual particles.

Since scientific jargon is the means of communication between the scientists of a specialized field and since it is very useful from a practical point of view, its importance is easily overestimated. However, in striking contrast to the scientific language, it could in principle be dispensed with. It is interesting to note that the research papers of the great scientists normally contain little or no scientific jargon, the papers of Einstein being an excellent example.

It is somewhat astonishing that Kuhn pays so little attention to the universal language of modern science, namely mathematics. Admittedly the degree of mathematical rigour varies considerably in different disciplines, but nevertheless the decisive role of mathematics is universally acknowledged. Also very often scientists use mathematical tools that are not established with full rigour, but this is not because they think mathematical rigour to be unnecessary, but only because they want to move ahead with their work while leaving the task of bringing rigour to formalism to the mathematicians. A classical example is the so-called δ function which was first introduced in electrical engineering and only brought into a rigorous form in the theory of distributions in the middle of the 20th century. The mathematical formalism quite often allows a continuous transition between the so-called "incommensurable" theories, as in the example of Newtonian and relativistic mechanics.

But even if the transition does not become fully continuous in the formalism, it is in many cases absolutely necessary to follow its development in order to understand the course of scientific progress. Viewed as if it was one of those things which Kuhn (1962) calls a "paradigm", there are several missing links in Kuhn's picture of classical mechanics. When Kuhn says that the paradigm of Newtonian mechanics was largely closed with Newton's "Principia", he misses completely the important role the formalism of analytical mechanics played from the 18th to the 20th century. The crucial links which were essential in establishing classical mechanics are notably (i) the important necessary additions which other scientists made, e.g., d'Alembert Principle; and (ii) the important aspect of the development of the formalism (by Euler, Lagrange, Jakobi, Hamilton, and Hertz). It is hardly conceivable that quantum mechanics and quantum field theory could have been formulated without this formalism, and indeed "the quantization rules" are framed in terms of entities introduced in this formalism of analytical, not original Newtonian, mechanics. Not only did this pave the way for an axiomatic formulation of quantum mechanics, the already established "canonical quantization rules" established there helped in the quantization of classical field theory. In a very short time, this led to the formulation of quantum field theory - up to now the most general form of quantum physics.

2.2.5   The Tree of Evolution: Interface-Building across Disciplines

A major difference between the methodological principle of dynamical core-context of scientific development advocated by us and the palaeontological "evolution tree" of Kuhn (1991a, pp. 3-13) has to do with the fundamental role played by the formation of interfaces between different fields of science (see the Fig. 2.1 below, contrasting Kuhn's tree of evolution with our methodological model of the core-context of development). As an example of the important contribution the dynamical core-context of scientific development makes in the creation of new fields, consider the case of the development of theoretical chemistry from classical (experimental) chemistry and quantum mechanics. An example of an interface creating a new field with great momentum is to be found in astro-particle physics. This is more than a mere merger, since the newly developing field has also its feedback on other fields. Astro-particle physics, e.g., allows scientists to ask research questions and pose problems in pure particle physics which could never be asked without the existence of the new field of astro-particle physics.

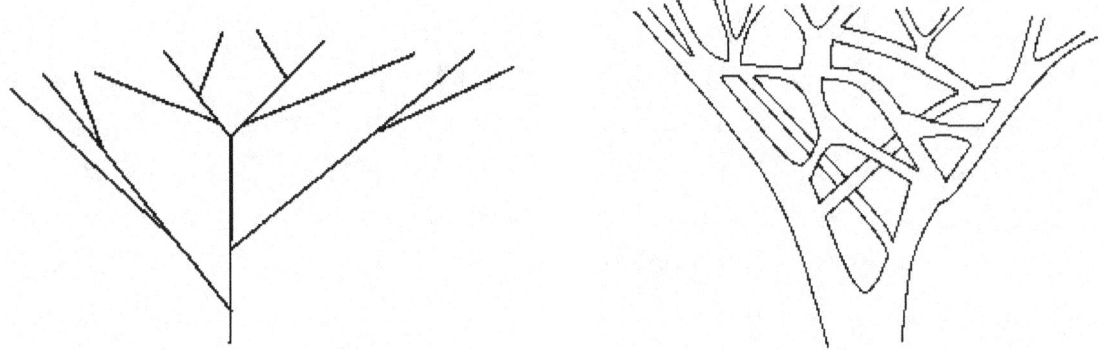

Figure 2.1: Left: A schematic view of the biological evolutionary tree of Kuhn (1991a, pp. 3-13). Right: Our picture representing methodological core-context building interfaces, in which theory-development and unification across the frontiers of fundamental physical theories play a fundamental role.

It is clear that the creation of such a new field is only possible if there is a sufficiently intensive communication between the respective scientific communities. This in turn is possible because the scientific language is universal - in contrast to the scientific jargon. Therefore, starting from the universal language, it is quite easy for a scientist to become familiar with a new field.

Another important interface-building between completely different branches of particle physics is due to a similarity in formalism. A very instructive example of this is particle physics and solid state physics. Here the subjects of the two fields are completely different. Solid state physics treats objects on the atomic scale. That is to say, it treats objects on a scale of nanometers, whereas particle physics treats sub-nuclear structures on a scale which is a million times smaller. Yet the formalisms of the two fields show striking parallels between them. Therefore, it is no surprise that scientists trained in one field could make seminal contributions to the other, the Nobel Prize winner R.

Wilson being an outstanding example. For lack of space we leave out an account of the interface building between theory and experiment at the active frontiers of physics (cf. Kane, G. & Pierce, A. eds.2008; and Pierce, A. 2008, pp.13-23).

2.3 Core-Contextual Methodological Structuralism

In the twentieth century meta-theoretical analytical tradition, a methodology of science is generally conceived as a normative philosophical doctrine about method (see Popper 1934) which may or may not bear a connection with actual scientific practice or with the foundations of science. If a methodology is normative, laying down rules regarding how the physicists ought to deal with a scientific theory, it is difficult to blame it for its ineffectiveness in physics, more so when we are concerned with issues in theory change by theory development. It is, therefore, hardly surprising, if we find that a dynamically conceived methodology of inquiry and theory-building in the sciences of physics and astronomy is missing from the methodological discourse of the past century. The task of such a methodology is to debate the question:

> How do the theoretical framework, the method, the core-context of theory development, the theory-experiment interface and the frontier of research questions co-evolve in these sciences?

Thus, the kind of methodology which we are here hinting at has the structural simplicity and dynamical complexity of physical theory at its core.

The question may be asked what is it which sets the methodology of dynamic core-contexts of theory development and problem-determination (Pandit 2002a, 2002b, 1991, 1982) apart from other methodologies, which have been developed by the philosophers in the past century (e.g., by Popper 1959, 1972a, 1972b). An important distinction between them lies in the fact that the latter have been invariably conceived as methodologies of theory-appraisal and theory-choice. These methodologies have been developed from a perspective which is taken from outside science, giving priority to the aim of science.

The aim of science is, in its turn, conceived in purely philosophical terms (e.g., by Popper 1934, 1963, 1972a, 1972b and others including Watkins 1984, Maxwell 1984, 1989, 2004, 2005, 2006, 2009 and van Fraassen 1980, 1987). Notice that while Watkins (1984) argues for a neo-Popperian philosophy of science, the choice of an optimum aim for science plays a crucial role in Watkins's attempt to improve upon Popper's falsificationism and critical rationalism (Pandit 1986, 141-148).

Moreover, as we have shown, they are not concerned with theory-development at all. Thus, the methodology of dynamic core-contexts of theory development is distinguished by *its* dynamic approach to the questions of structural identity and scientific change *within* physics. This being a major difference, its basic principle, viz., core-contextual methodological structuralism, emphasizes *two* things in the methodology of physics. One of these concerns the *inner development* in physics, in terms of both theory-development and problem-determination. The other concerns the guidance the methodology of physics itself receives at its active frontiers from this *inner development*. Besides this, methodological structuralism entails strong correlations between the core-contextual theory development and evaluation of a physical theory (see Chapter 8).

The questions of structural identity and scientific change within physics can accordingly be treated by specifying the *correlated* core-context of theory development and the frontier of research. Such specification is possible by means of a *two-step* procedure. *First*, it is necessary to identify a physical theory T (possibly a set of theories) as a mature and successful dynamic core-context of theory development and the underlying fundamental principles, the formalism and its relation to observables. In this step, it is important to specify, *over and above* its successful record of performance by virtue of its explanatory power, the interesting and new probing research questions which T poses by virtue of its *resolving power*. In the *second step*, it should be possible to specify the kind of new (*possibly unifying*) theory T' which physics at its open frontiers *can* aim at, *given* the core-context of theories participating in this development. Thus, to think of the *core-context of development* in physics is to think about those physical theories that fulfil at least the following three conditions:

(C1) they not only explain all that which they are expected to explain but decide what can be observed or measured, given the state of the art of instrumentation and experimentation at the present and foreseeable future; they will also strongly influence the techniques of experimentation;

(C2) being highly successful, they participate in theory-development and, *possibly*, in the methodology of theory-unification; *and*

(C3) being sufficiently rich, mature and powerful, they pose new probing research questions and create *new* frontiers of discovery, making physics open-ended.

All these properties are clearly outside the scope of a purely logical analysis of the language of science. Obviously, these can be treated only inside the dynamic core-context of physical theory. In the proposed methodology a close *connection* between the core-context guided theory-development and theory-unification cannot be missed. While no theory-unification (involving well-established/articulated theories) without the development of a new theory is possible, not every case of theory-development need necessarily end as a case of theory-unification. Thus, theory-unification, which entails theory-development, must also receive guidance from the core-context of theory-development.

The core-contexts of scientific development that are *internal* to physics and astronomy and, *therefore*, methodologically important to their frontiers of research have received in the philosophy of science neither the recognition nor the attention which they deserve. This is more glaringly true of the following question which has remained unasked:

> Whether, or how, a methodology of theory-development in physics is possible which can at the same time be considered as a methodology of theory-experiment interface building and theory-testing?

In order to develop such a methodology we have to avoid the shortcomings of the purely meta-theoretical approach. Together with the methodological tools already expounded in sect. 2.1.3, we now want to build on this basis the methodology of dynamic core-contexts of development. We first reconsider the two dogmas of exclusion that have dominated the twentieth century philosophy of science, which have been expounded in sect. 2.1 above. Found in different versions, the first dogma of exclusion originally arose as a universally accepted answer to the following question, assuming a sharp distinction between the context of scientific rationality and the context of scientific discovery: *How should philosophy of science articulate scientific rationality?*

Conceived as a context in which the scientist tests, justifies, or appraises and validates theories depending upon how they relate to bodies of experimental evidence, the former but not the latter is held to be exclusively relevant for philosophical or logical analysis of completed scientific theories. The aim of logical analysis is the reconstruction of the context of *logical relations of deduction* between the theory and reports of observation or experiment, on which the scientist depends for the decision either to accept or reject the theory. On the other hand, the context of scientific discovery is conceived as a context of many factors or influences which are *external* to science. Although these do not build on any logical relations, they decisively shape science from outside, in so many ways which may be of interest to policy makers, to sociologists, economists and psychologists. The context of scientific discovery thus *excluded* from scientific rationality is the same as *science in context* excluded from philosophy of physics and science. As a corollary of the first *dogma of exclusion, the second dogma of exclusion* arose originally as follows: There is no connection between *theory-testing* and *theory-forming* in physics. The former is governed by deductive logic but the latter by the process of discovery which is itself governed by psychological factors. In the absence of a connection between them, there is nothing to warrant a study of the one which is at the same time a study of the other.

At least Norwood Russell Hanson (1961, 1965, p.3)[13] together with other thinkers (notably Peirce) before him explicitly articulated this view of science, clearly implying a separation between theory-testing and theory-forming. Consequently, Hanson, a staunch advocate of *the logic of discovery* in science, completely missed the *core-contexts of theory-development in physics*.[14] One of the main reasons being this:

---

[13]To quote Norwood Russell Hanson (1965, p.3): "The issue is not theory-using, but theory-finding; my concern is not with the testing of hypotheses, but with their discovery".

[14]This may look rather strange because Hanson was among the very few great theorists of science whose deeper philosophical interest in the logic of scientific discovery remains unsurpassed, with the exception of some methodologists of physics who were also the practicing physicists. Pierre Duhem (1906) is the most leading and notable among them all.

With his attention focused on the logic of discovery, Hanson (1961, pp. 21-23) did not break with the strong logical tradition of analytic philosophy of science. In other words, as is true of the logical empiricist Carnap, Hanson too stayed very close to the tradition of understanding science from outside, relying mainly on the logical tools of meta-theory. This is amply clear from Hanson's (1961, p. 22) statement elaborating his distinction between "reasons for accepting a hypothesis H" from "reasons for suggesting H in the first place":

> What would be our reasons for accepting H? These will be those we might have for thinking H true. But the reasons for suggesting H originally, or for formulating H in one way rather than another, may not be those one requires before thinking H true. They are, rather, those reasons which make H a plausible type of conjecture.

In admiration of Hanson's analysis, the philosopher of science Wesley C. Salmon (1966/ 1967, pp.111-115) even suggested an improvement by recognizing three logically distinct aspects of treatment of a scientific hypothesis H: (1) thinking of H; (2) plausibility considerations for proposing H; and (3) testing or confirmation of H. Salmon (1966/1967, p. 114) argues: '

> Hanson has argued (correctly I think) that there is an important logical distinction between plausibility arguments and the testing of hypotheses, but he has (mistakenly I think) conflated plausibility arguments with discovery.

The status of plausibility considerations themselves seems, however, debatable, as is indicated by the following question which Salmon (1966/1967, p. 114) asks:

> Are plausibility considerations psychological or subjective in character? Do they play a legitimate role in science, or do they merely express the prejudices of the scientist or the scientific community? Are they different in kind from the considerations involved in the confirmation of hypotheses? An answer to this question will be forthcoming when we look more closely at what the probability calculus tells us about confirmation.

Nevertheless, we think that Hanson (1961, p. 23) opened a line of inquiry in the right direction, as in the following instance, even though it exclusively focused on the logic of discovery:

> The issue is whether, before having hit a hypothesis which succeeds in its predictions, one can have good reasons for anticipating that the hypothesis will be one of some particular kind. Could Kepler, for example, have had good reasons, before his elliptical-orbit hypothesis was established, for supposing that the successful hypothesis concerning Mars' orbit would be of the noncircular kind?

It is beyond the scope of the present study to go deeper into Hanson's meta-theoretical problem of the logic of discovery, conceived as the problem of determining which conditions are relevant for assigning a high prior probability to a hypothesis H in order to categorize it as belonging to the class of plausible conjectures, as distinguished from its posterior probability which "yields a *posterior weight* which is based upon plausibility considerations *and* confirmatory evidence"(see Salmon, 1966/1967, p.124-130). We do not deny the importance of the logical task of analysing Hanson's problem by reducing it to the logical problem of identifying criteria relevant to the prior probabilities of scientific hypotheses. But all this is not crucial to our approach or to our argument in the present study.

In sharp contrast to (Hanson's project of) the logic of discovery, the concept of *the dynamic core-context of theory development* emerges as a very powerful concept in so far as it plays a central role in the methodology of theory development and unification. The present study focuses on all the above stated conditions (C1 - C3) of theory-development. We illustrate these in terms of developments in particle physics leading to its Standard Model and possibly beyond it. As a *core-context of development*, the Standard Model is highly successful yet confronted with *new* probing research questions which it would not have been possible to pose in absence of its own framework. Thus, it has opened a new frontier of research questions with the possibility of finding a more comprehensive theory which goes beyond it without necessarily contradicting its achievements. This clearly indicates the fundamental role it plays at present as the *core-context of theory development* in particle physics, not only in theory-testing and theory-finding but in theory-experiment interface building. Equipped with such a rich key-concept as the *core-context of theory development*, in what follows our discussion aims at showing how the methodology of dynamic core-context building, and re-building, fares as a methodology of theory-finding and theory-testing in physics. The dynamic core-context constantly builds and re-builds interfaces across the successful theories, even shaping the state of the art of instrumentation and experimentation. The physicist active at the frontier situates himself in it. On the one hand, if he does so, it is to *reflect* on the new probing research questions. On the other hand, he searches for the kind of fundamental theory which physics can aim at, testing the possible candidates as far as possible. *Thus, although the new theory is beyond the reach of the core-context, it will nevertheless be shaped by it.* As one of its great advantages, the proposed methodology enables us to reconstruct coherently theory-development and unification across physics. Its great simplicity can be demonstrated by deriving Duhem's methodological holism (Duhem 1906) concerning theory-testing as one of its consequences (Pandit: 2002a, 2002b): In our interpretation *then*

> Duhem was dealing with one particular aspect of context-building when he so perceptively related it to theory-testing by arguing that it is always a cluster of hypotheses H' (= H and A) which face a test, but never the isolated hypothesis *h* which we all the time think of as the candidate for the test. In a nutshell, in theory testing, the candidate theory is not free from the context of theories participating in theory development. Once we recognize the role of context-building in theory-testing in Duhem's sense, it is not possible to escape or ignore its implications for the interpretation of the results of theory-testing, particularly with respect to the methodology of theory-choice.

In this way, the methodology we are proposing can be shown as the methodology of dynamic core-context building in theory-finding and theory-testing. It enables us to have a fresh look at the entire structure of inter-theoretical relations in physics. A fundamental principle of this methodology which we propose to discuss here (turn to sect. 4.5 below) with reference to some of the best known attempts at unification is that physics and astronomy develop by core-context re-building marked by successful unifications.

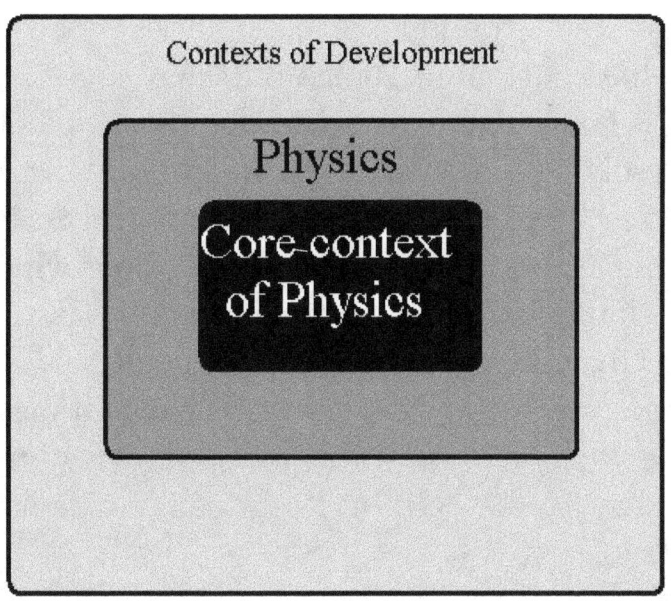

Figure 2.2: A schematic view of the dynamic core-context of theory development which is internal to physics, distinguishable from the contexts of development which are external to physics.

Another fundamental aspect of a physical theory/core-context which methodological structuralism emphasises is its resolving power. It is by virtue of its resolving power that a theory is able to determine new probing research questions, even questions that prepare the ground for its participation in the search for a better alternative to it. Be it theory-development or theory-unification, the core-context of theory-development, consisting of one single unified theory or several theories, will be known, on the one hand, by its record of *consistently successful explanations* of the experimental data and, on the other hand, by its *resolving power. The resolving power of a physical theory refers to its power to pose new interesting and probing research questions which were not thought of before* (Pandit, 2002a; Pandit 1982: 12-14, 91-92, 96,110; Pandit 1991: 212-213, 304-306, 309-314, 354-355). If we want to know about the new frontier with a new horizon which a physical theory opens up in terms of these *new* research questions, we can do so only with reference to the *core-context of theory-development.* As a rule, a physical theory, embedded in the core-context, points by virtue of its own successful performance to the frontier of research beyond its own regime of applicability. Clearly, the questions of what can and what cannot be observed, measured and tested as per above conditions C1 - C3, given the theory and the predictions it makes, will always depend on the state of the art of instrumentation and experimental possibilities as an essential input from the *core-context* (Pandit 2002a: 361-369).

What does it really signify, in scientific terms, to look at a physical theory, with resolving power, as *the dynamic core-context of development*? It signifies a serious comparison of physical theories on a scale of lower and higher resolving powers. It is these which prepare the physicist for asking new probing research questions that can open a new frontier. Given a physical theory T in its field, with a certain resolving power, we can lay down a set of conditions for the *inner development of a physical theory* in terms of a new theory T', with a higher *resolving power*. Given T, we can say that T' has higher resolving power only if

> T' poses interesting and new probing research questions, possibly beyond the reach of T such that (C4) in comparison to the research questions posed by T, there are new research questions posed by T', which far *exceed* (i) in number (ii) in depth (iii) in creating a new frontier, both in terms of the future possibilities of theory development and new frontiers of instrumentation and experiment; *and* (iv) in exposing T' itself, by virtue of its qualitatively novel predictions, to higher degrees of testability.

For best examples, we may turn to quantum field theory and Einstein's special and general theories of relativity. The interaction between quantum theory and the special theory of relativity necessitated the development of relativistic quantum field theory, as a new *core-context*. A similar interaction between quantum field theory and the classical field theory of gravitation in general relativity may give rise to new theories. Where potential candidates are in sight, they are yet to be brought to maturity and tested experimentally, if possible.

Scientifically important physical theories predict the existence of hitherto unknown and unobserved phenomena, going beyond what the scientists can expect to observe or know on the basis of the well-established old theories. They necessitate the setting up of new experiments and instrumentation of very high sensitivity to make appropriate measurements with a view to confirming whether or not the kind of phenomena which are predicted really exist. Sometimes the new experiments are only possible by the development of a new core-context in another field. A very good example is the development of radio-astronomy on the one hand and precision tests of general relativity on the other. An example of those precision tests within the known core-context is the tracing of space crafts in the solar system. The Pioneer 10-11 space craft, the ones most deeply penetrating in outer space show indeed an anomaly which seems in contradiction with the theory of general relativity as it stands (Anderson et al 1998, 2858). At the moment there are several attempts to explain this anomaly within the current core-contexts of gravity and field theory. There is, however, a (perhaps remote) possibility that this anomaly plays a similar role as was played by the perihelion precession of Mercury. It might not be the cause of a major change in the current core-context, but it might fit neatly into a newly developed one.

Radio-astronomy was the offspring of the development of physics *in* context, the context comprising of the development of radar in World War II. It is with this method that the pulsars were detected. The acceleration of their period due to the emission of gravitational waves, which is very precisely predicted by the theory of general relativity, is now one of the, if not the most precise and conclusive, tests of that theory. The experiments (e.g., of Hulse and Taylor 1975, 1993) "show that, whatever the precise directions of future theoretical work may be, the correct theory of gravity must make predictions that are asymptotically close to those of general relativity over a vast range of classical circumstances" (Taylor, Jr., J. H. 1993, p. 89).

Thus, without the right kind of theory being in place, a physicist would not even know where to look, for *possibly* new type of phenomena (see Valtonen et al. 2008). These new phenomena are in one of their aspects considered as a "final proof" of the new theory, e.g., electromagnetic waves in Maxwell's theory of electromagnetism and the existence of anti-matter in Dirac's relativistic quantum physics. But this is only one aspect, since the new phenomena are quite often the source of new and serious problems. These new phenomena and new problems offer sometimes exciting new fields in physics and even technology (e.g., think of radio-waves in electrodynamics). But very often they also give rise to important difficulties in applying or even formulating the new theory. A very good example in this respect is the existence of anti-matter. On the one hand, it leads to fascinating new theoretical and experimental possibilities, reaching even into medical diagnosis, the Positron Emission Tomography (PET) being a good example. On the other hand, because of the possibility of virtual pair creation, it is the source of fundamental difficulties in relativistic quantum field theory. These difficulties have been solved only partially by the so-called renormalized perturbation theory of relativistic quantum field theory (see e.g., Dosch et al 2005b). This is very illustrative of how physical theories can be structurally simple and yet dynamically complex.

It is remarkable that often a discovery being made on the basis of the predictions of a particular theory discloses new type of phenomena that are completely unsuspected. In an over-simple meta-theoretical approach to physical theory, these new phenomena are regarded as potential falsifications of the theory so long as they are not observed experimentally. Dark matter is an example.[15] It has been observed for quite some time that the rotation of galaxies is not in accordance with Newtonian gravity. Without proper scientific reasoning based on the core-context guided inner development of physics, it would have been an obvious consequence that the theory of gravity has to be modified. Perhaps in the beginning of the 19th century this would have been the first choice. But in the middle of the 20th century, theory of gravity including general relativity was so much a part of the core-context,

---

[15] It was Fritz Zwicky (1933) who made observations of the Coma Cluster that indicated the possible presence of Dark Matter. The frontier of Dark Matter was slowly unfolding here as a simple consequence of the measurement of mass in the core-context of the relevant theory of gravitation and velocity measurements. Employing arguments that build upon simple kinematics "Zwicky realized that the numerous galaxies that made their home in the cluster were moving too fast. The gravitational pull of the luminous matter present in the cluster was insufficient to balance the outward 'centrifugal force' of the matter. Without a dark component, the cluster would pull itself apart" (Aaron Pierce, 2008, pp. 13-14). Applying analogous kinematic arguments in the case of the rotation of individual galaxies, astronomers have arrived at the conclusion that galaxies possess a halo of Dark Matter. Here the arguments relying on simple kinematics build up as follows: "The rotation speeds of galaxies themselves can be determined as a function of radius by making red-shift measurements, and taking advantage of the optical Doppler effect. Newton's law of gravitation indicates the rate of rotation depends on the enclosed mass. If luminous matter were the whole story, then the rotational speeds of galaxies should begin to fall off with radius beyond the bright core as $v(r) \sim r$. This is not what is observed - rotation curves remain stubbornly constant at radii far exceeding the luminous core, indicating a dark component of mass. This evidence supports the notion that galaxies possess a Halo of dark Matter". For a simple account of the alternative measurements applying the kinematic argument to the regions of space representing a fair sample of the universe as a whole and yielding consistent results that are indicative of exotic Dark Matter.

that one took a different path altogether. One assumed the existence of dark matter. In the meantime further scientific reasoning reveals that there are many more physical indications for its existence. Combining relevant observation with other theories, scientific reasoning leads to the much more "revolutionary" result that most of the dark matter cannot be of a conventional nature. Once these phenomena are discovered, their precise and detailed theoretical identifications made and the relevant parameters measured, they will open up unexpected new frontiers full of new probing research questions before our eyes, taking the physicist and his work by surprise.

More than a century ago, it is remarkable how aptly J. C. Maxwell (1890) hinted at this inner development of physical theory in the following words:

> From the straight line of Euclid to the lines of force of Faraday this has been the character of the ideas by which science has been advanced, and by the free use of dynamical as well as geometrical ideas we may hope for a further advance ... We are probably ignorant even of the name of the science which will be developed out of the materials we are now collecting ...

2.4 Principle theories and constructive theories in physics

Physics is not distinguished by theory-building and theory-testing algorithmically. On the contrary, it is distinguished by creative theorizing. This is the standard view held by the most leading physicists of the 20th century, notably by Albert Einstein (1934, 116) and others. To quote Einstein:

> "A new idea comes suddenly and in a rather intuitive way. That means that it is not reached by conscious logical conclusions. But thinking it through afterwards you can always discover the reasons which have led you unconsciously to your guess and you will find a logical way to justify it".[16]

---

[16]Einstein to Dr. H. L. Gordon, 3 May 1949 (Item 58-217 in the Control Index to the Einstein Archive) as cited in Stachel, J. (1994, p.146).

This view, which is quite consistent with the creative role mathematics plays in physics, does not imply that creative theorizing by the physicist gives him total freedom to speculate wildly. We want to argue for the view that this creative theorizing must take place within the dynamic core-context building set by the core-theory, i.e., within the dynamic core-context building and re-building in physics. Interestingly enough, Duhem's (1906) holism concerning theory-testing (considered in sect. 2.3 above), sometimes called holistic conventionalism, is quite compatible with our view.

Einstein was one of the very few physicists (Ernst Mach, P. Duhem, Poincare and Werner Heisenberg being other examples) whose contribution to theory development is matched by their contribution to methodology and foundations of physics. This becomes so clear if we ask the question whether the physics community today is not a fragmented community, given that not every physicist currently working at the frontier of theory development in search of a theory beyond the Standard Model of particle physics takes interest in the foundational and methodological problems. Einstein (1934: 127) has himself drawn a methodological distinction between two kinds of theories in physics, i.e., between (1) principle theories and (2) constructive theories, as he called them. Principle theories are seen by him as a necessary step to constructive theories, the former exercising significant methodological constraints on the latter. Of great foundational significance as this is to physics, we can look at quantum field theory as a principle theory and at Standard Model of particle physics as constructive theory. The latter would not have been possible without the former as its fundamental theoretical and mathematical framework. This distinction provides an important insight into theory development in physics. It is an important reminder how one and the same physical theory which is simple structurally could be complex dynamically. It may take a long time before a physical theory, which is simple structurally (Einstein's general theory of relativity being a good example), reveals its dynamic complexity, depending upon the core-context, the theory-experiment interface, the technology of instrumentation and experimental framework within which its consequences develop over a period of time.

Einstein characterises the two kinds of theories as follows. Being the most important class of theories, the constructive theories "attempt to build up a picture of the more complex phenomena out of the materials of a relatively simple formal scheme from which they start out. Thus the kinetic theory of gases seeks to reduce mechanical, thermal, and diffusional processes to movements of molecules - i.e., to build them up out of the hypothesis of molecular motion. When we say that we have succeeded in understanding a group of natural processes, we invariably mean that a constructive theory has been found which covers the processes in question" (Einstein 1934, 127).

As regards the *principle theories*, they "employ the analytic, not the synthetic method. The elements which form their basis and starting point are not hypothetically constructed but empirically discovered ones, general characteristics of natural processes, principles that give rise to mathematically formulated criteria which the separate processes or the theoretical representations have to satisfy. Thus the science of thermodynamics seeks by analytical means to deduce necessary conditions, which separate events have to satisfy, from the principle that perpetual motion is impossible" (Einstein 1934, 127-128). Einstein (1934, 128) then points out: "The advantages of the constructive theory are completeness, adaptability, and clearness; those of the principle theory are logical perfection and security of the foundations". In view of its logical completeness, if a single consequence from among all the consequences that are drawn from *a principle theory* "proves wrong, it must be given up; to modify it without destroying the whole structure seems to be impossible" (Einstein 1934, 131). This may also be seen as an indicator of its simplicity. The theory of relativity, regarded as *a principle theory* (Einstein 1934, 128) is likened by Einstein to a building consisting of two separate stories, the special theory and the general theory.

As regards the *principle theories*, one might like to think that a logical scheme independent of the core-context is what suits them best. That this is not the case can be clearly seen by the role the so-called cosmological constant played in general relativity. The original Einstein equations did not allow for a static solution of the universe, which at the time was the Standard model of cosmology. Therefore Einstein added a term to the equations, which was not forbidden by general coordinate covariance, but which did not follow from the inner logic of Einstein's theory. Shortly after Einstein had introduced this term, new experimental evidence obtained by Hubble (1929)[17] indicated an

---

[17]Hubble (1929) investigated the relationship between distance and radial velocity of extra-galactic nebulae. His findings revolutioized astronomy, palying a fundamenbtal role in opening up new frontiers of research.

expanding and not a static universe. Einstein regretted to have introduced the cosmological constant and called it the biggest blunder of his life. This looks like a triumph of logic over "extra-logical considerations", but it was not the end of the story. Relativistic quantum field theory made it plausible, albeit not necessary, to introduce the cosmological constant again. But the cosmological constant was so large that a resultant universe had to be very small.[18] There is yet no way in sight to combine the principles of general relativity with those of quantum theory and therefore the problem of the cosmological constant leads not to a logical contradiction. Most physicists working in particle physics or cosmology consider it, however, as the most important problem of the field, if not of fundamental physics at all.

It would be a misunderstanding, if one took Einstein's distinction between the principle theories and constructive theories dogmatically. Indeed, by the time Einstein had drawn this distinction his own views regarding the creative aspect of physics had undergone a tremendous change. On the one hand, his views had moved closer to the idea that "the creative principle resides in mathematics" but, on the other hand, away from his earlier strong inclination towards radical empiricism. In a lecture on general relativity which he (Einstein 1934, p. 48) delivered in London in the year 1921, he said:

> "I am anxious to draw attention to the fact that this theory
> is not speculative in origin; it owes its invention entirely to
> the desire to make physical theory fit fact as well as possi-
> ble. We have here no revolutionary act ... The abandonment
> of a certain concept ... must not be regarded as arbitrary,
> but only as conditioned by observed facts ... the justifica-
> tion for a physical concept lies exclusively in its clear and
> unambiguous relation to facts that can be experienced. ...
> The general theory of relativity owes its existence ... to the
> empirical fact of the numerical equality of the inertial and
> gravitational masses of bodies. ... "

However, later on in the second lecture, delivered at Oxford in 1933, while dealing with the relation between experience and reason in physics, Einstein (1934, p. 116) asked:

---

[18]Even Switzerland would not fit in such a tiny universe, remarked W. Veltmann in an official speech delivered at Geneva, Switzerland.

"If, then, it is true that this axiomatic basis of theoretical physics cannot be extracted from experience, but must be freely invented, can we ever hope to find the right way? ... I am convinced that we can discover by means of purely mathematical constructions the concepts and the laws connecting them with each other, which furnish the key to the understanding of natural phenomena. ... the creative principle resides in mathematics".

This change of attitude might have been nourished by the observation that the principle of general covariance, an important cornerstone of the general theory of relativity, is devoid of physical content (Kretschmann 1917, 575). Though this seems from outside a catastrophe, it is not. The theory of general relativity had in the meantime gained enough dynamical structure to bear this blow. Einstein seemed to have granted from then on only the physical meaning of a heuristic concept. It should be noted that even today there is no general agreement on the definition of "background independence", generally thought to replace the principle of general coordinate invariance. This does not impede progress in gravitational theory. What is important is that in the context *in* physics one can give this principle a well-defined mathematical meaning (see Giulini et al 2007, pp. 105-120).

What is important from the point of view of methodological structuralism is this: Einstein's distinction does not suggest or imply that theory development either in the sense of *the constructive theories or the principle theories* is free from any guidance or influence from *the core-context of development* within physics.

To conclude our discussion, we can say three things about physical theory as follows, which are methodologically important: *First, all* theories, whether constructive theories or principle theories, carry within themselves the guiding influence of the core-context of development. *Secondly,* their dynamic complexity is such that they do not indicate that it is possible at all to separate the research *questions of existence* of the newly predicted particles and the *questions of the theoretical framework*, within which the former must be pursued, into two distinct categories. Quite evidently, correct answers to *the research questions of existence* of newly predicted particles in physics have direct implications for questions concerning the correctness of their theoretical framework. And *thirdly,* as we have seen in sect. 2.1.3 above, "It is always the theory which decides what can be observed" (Heisenberg 1983, 1984, 1985, Pandit 2002a, 2002b).

Chapter 3

THREE CASE STUDIES IN PHYSICS

Wenn wir, statt den Theorien über die Physik, vielmehr dem Prozess der physikalischen Begriffsbildung selbst folgen, so werden wir vielleicht erwarten und hoffen dürfen, in ihm gewisse Grundbestimmungen aufzufinden, die gegenüber dem Wechsel der verschiedenen erkenntnistheoretischen Bezugssysteme invariant sind.
Cassirer 1937, p. 38

3.1 Introduction

We present three case studies. Each of these is intended to illustrate the criticism and the programme expounded in the Chapter 2. The *first two* cases, "Helmholtz's Weakest Work" and Yukawa's Meson Theory, pertain to our thesis that in the philosophy and methodology of science a purely logical/meta-theoretical analysis is bound to be inadequate and that the dynamic aspects such as the core-context of development and the resolving power of physical theory have to be taken into account. Otherwise, it is not possible to understand theory change by theory development in physics in its full dynamic complexity. The *third case*, the $\theta - \tau$ puzzle, shows the asymmetry between theories and problems. It also shows how the solution of puzzles in physics can create problems which open up new frontiers full of new probing research questions.

3.2 "Helmholtz's Weakest Work"

"Helmholtz's Weakest Work" is the title of a short section in a book on the history of Thermodynamics by C. Truesdell(1980). A short discussion of it can serve as an illuminating illustration of the differences between an outside and an inside view of science, i.e. the difference between an analysis in the sense of the meta-theoretical logic and one in terms of the inner dynamics of science, as it is called here.

Truesdell refers to the paper "Über die Erhaltung der Kraft" by the young military surgeon H. Helmholtz (1847). That this paper has its shortcomings is evident. But this was already known for a long time. The critique of Truesdell is somewhat schoolmasterly petty but correct.

Notwithstanding its apparent shortcomings, this paper is generally considered as one of the four seminal works leading to one of the most important principles of physics, namely the principle of energy conservation.[1] One of the greatest physicists, James Clerk Maxwell (1877, p. 389) writes about this allegedly weak paper:[2] "There is no doubt that a very great impulse was communicated to this research[3] by the publication in 1847 of Helmholtz's essay ..." And more specifically he adds: "To appreciate the full scientific value of Helmholtz's little essay on this subject, we should have to ask those to whom we owe the greatest discoveries in thermodynamics and other branches of modern physics how many times they have read it over, and how often during their researches they felt the weighty statements of Helmholtz acting on their minds like an irresistible driving power". We may assume that Maxwell, to whom we owe indeed the greatest discoveries in modern physics, belonged among the readers.

Equally enlightening are the remarks of Helmholtz himself on that paper. On the height of his fame at a celebration in his honour[4] he remarked that as a student (of medicin) he had the time to read in his spare time the works of D. Bernoulli, D'Alembert and other mathematicians. In his own words (Helmholtz, 1892, p. 51):

> "So the question came to my mind: 'Which relations have to exist between different forces of nature if a perpetuum mobile must be impossible?' and the further one: 'Exist these relations indeed?'. It was my purpose to give in my

---

[1]See, e.g. Kuhn (1959). The three other persons involved are Mayer, Joule and Colding.

[2]J. Clerk Maxwell, *Nature* XV (1877) p. 389, Papers, vol. 2, p. 592

[3]That is, to the establishment of the principle of the conservation of energy

[4]It was the celebration of the 49th anniversary of his promotion to a doctor of medicine. The addresses are collected in a small booklet (Helmholtz 1892): *Ansprachen und Reden ... zu Ehren von Hermann von Helmholtz*, Hirschwald'sche Buchhandlung: Berlin 1892.

booklet on the conservation of force ⟨ that is energy⟩ only a critical investigation and establish an order of the facts in the interest of physiologists.

I would have been fully prepared for objections of the experts like 'That is all well known. What goes on in the mind of the young medical doctor that he thinks that he has to represent that to us in such a detail? To my astonishment the physical authorities ... received that in a completely different manner. They were inclined to deny the validity of the law ⟨of energy conservation⟩ and, engaged in their fight against the natural philosophy of Hegel, to declare my work as a fantastic speculation, too. Only the mathematician Jakobi recognized the connection of my thoughts with the mathematicians of the past ⟨18th⟩ century..."'.

Thus, in the words of the two eminent scientists, Maxwell and Helmholtz, we find an excellent description of what we have here proposed to call the core-context of development, bringing out the importance of the principle of the resolving power of a theory. Notice that there is at the same time also a search for theory unification and for the creation of new frontiers. Puzzles might be solved even inadequately, but this might lead to new frontiers with new problems. The (very civilized) opposition of the physical authorities of Berlin at that time[5] indicates that there was a new frontier opening up, but Jakobi,[6] concentrating on the form and methodological aspects, saw clearly that there was also an underlying continuity. On the other hand, Truesdell in his criticism concentrates only on the purely logical aspects of the paper and, indeed, it is hardly surprising if a meta-theoretical analysis in his style would have resulted in a rather low rating. The enormous resolving power of the paper is however totally ignored or missed.

Significantly enough, it was Pauli who classified research papers in five categories according to whether the results were (i) *new and*

---

[5]Most prominently it was the professor of physics Magnus, who supported Helmholtz morally and financially. He objected to the publication of Helmholtz's work in the periodical *Annalen der Physik* but found an editor for him who published the booklet (and even paid a small honorarium).

[6]Jakobi and Hamilton were the two mathematicians who brought classical mechanics in a form which turned out later to be necessary to formulate the principles of quantization.

right (ii) *new and wrong* (iii) *old and right* (iv) *old and wrong*, and (v) the last and most deplorable category was: *not even wrong*. Clearly, Helmholtz's paper can be placed between the category one and two, pure logicians of science would presumably prefer papers from the fifth category.

3.3   Yukawa's Meson Theory

In this section, we consider the early development of meson theory, initiated by Yukawa in 1935 (Yukawa 1935, p. 48). It was based on relativistic quantum field theory and it predicted the existence of a new particle responsible for the forces which keep the atomic nucleus together. Though the new theory and the prediction of Yukawa was extremely bold, it nevertheless can be seen as part of the developmental process of unification and extension of the core-context which was already in place.

In 1935 relativistic quantum field theory was far from being a closed and self-consistent theory, but several theoretical and experimental developments lead to its gradual acceptance within the core context. Though the origins of quantum field theory go back to the time of the establishment of "new" quantum mechanics (Klein and Jordan 1927, p. 751), it was only the unification of quantum physics and special relativity which showed that a quantized field theory is indispensable. It is noteworthy that one of the principal predictions of relativistic quantum physics, namely the existence of antiparticles (Dirac 1930f, p. 60), was experimentally verified in 1932 (Anderson 1932, p. 238). Yukawa (1935, 189) writes that the year

> "1932 was even more turbulent for physics than it was for my own life.[7] Events, each of which, taken alone, could be called revolutionary occurred three in a row: first, there was the discovery of the neutron, second the discovery of the positron;[8] third the atomic nucleus was broken by artificial means ... The discipline that is now called nuclear physics was until then a minor branch of study, but because of those three events, it suddenly became the mainstream".

---

[7]He got married that year.
[8]That is the antiparticle of the electron.

The young lecturer Yukawa was until then mostly concerned with the seeming inconsistencies of quantum field theory, namely the infinite energies occurring there in certain expressions.[9] But in 1932 he turned his interest to this new field of nuclear physics. He wrote an introduction to Heisenberg's papers on nuclear physics in the *Proceedings of the Physico-Mathematical Society of Japan* and at the same time decided "to carry the theory one step further". He chose to tackle the problem of nuclear forces. By confronting this problem he committed himself to long days of suffering. In fact, he describes the period from 1932 to 1934 as the most difficult years of his life. Already at the beginning of his investigations it seemed likely to him that the nuclear force was a third fundamental force unrelated to gravitation and electromagnetism and that this nuclear force could also find an expression as a field. He reports:

> "I had this idea of a nuclear force field very early. Looked at from the quantum mechanical viewpoint, a field of force almost by necessity implies that there is a particle accompanying that field. ... Stated this way, the answer appeared almost at hand, but my brain did not work so quickly. I had to take the wrong path first, before I could arrive at my destination" (Yukawa 1982, p.194).

We see here in Yukawa's own words the opening of a new frontier with its own problems, the nuclear interaction, and at the same time the desire for unification, namely to describe the new interaction within the known frame, namely a (quantized) field theory.

He was brought on the right track by the core-context of development in a different field, namely that of weak interactions. In 1934 Fermi had developed a quantum field theory of radioactive beta-decay (Fermi 1934, p. 61), in which an essential feature of relativistic quantum field theories, namely the creation and annihilation of particles, plays an important role. Yukawa was wondering whether the problem of the strong nuclear force could not be solved by exchanging an electron and a neutrino between the nucleons, but he found out that other physicists[10] had been working on the same idea and their results were negative, the resulting force was much too small.

---

[9]He writes: "I read their report (a paper by Heisenberg and Pauli on quantum electrodynamics) many times and I considered thoughtfully every day how I could debit this devil called infinity. But the demon was stronger than I (p. 173)".

[10]Notably, these were mainly Heisenberg and Majorana.

So Yukawa went further: he adopted the frame of quantum field theory but he introduced a completely new field which mediated the nuclear interaction. The particle corresponding to this field was different from the matter known up to then, viz., the electrons, nucleons, and the (hypothetical) neutrinos. Yukawa realized that within the field theoretical framework he could predict the mass of the new particle from the range of the nuclear forces. It should be distinctly heavier than the electron and lighter than the proton and neutron; he estimated the mass to be around 200 electron masses. He presented his new theory to the Osaka branch of the Physico-Mathematical Society of Japan. Nishina, one of the most important Japanese physicists who, together with O. Klein, had written important papers on quantum electrodynamics, congratulated him and an English version of the paper was published in the same month of November 1934.

A. Pais (1986, p. 432) describes the paper as follows: "Yukawa's paper is a primitive improvisation, yet it contains three major lasting points: the dissociation of the strong nuclear force from the weak beta-decay force; a sensible first estimate of the meson mass; and the remark that 'the massive quanta may also have some bearing on the shower produced by cosmic rays' ". From a meta- theoretical point of view the paper is thus hardly noteworthy, but it contributed essentially to the dynamic core-context of particle physics and became a true cornerstone in the physics of elementary particles.

Yukawa's ideas remained for some time unnoticed, but this changed in 1937. In a detailed paper on cosmic rays, Neddermayer and Anderson (1937, p. 884) came to a far reaching conclusion: Either the standard theory (that is the quantum theory of electrodynamics) fails or "there exist particles of unit charge but with a mass ... larger than that of a normal free electron and much smaller than that of a proton". With this tentative establishment of a new particle the theory of Yukawa received attention. In the same year Oppenheimer and Serber (1937, p. 1113) proposed[11] that the newly detected particle is the one originally proposed by Yukawa in order to explain nuclear forces. The shortcomings of the theory were emended and the development culminated in the prediction of another new particle, this time a neutral one. The predictions (Kemmer 1938, p. 354) were based on symmetry arguments and introduced a new realm of application of symmetries, the so-called inner symmetries. In 1938 Oscar Klein

---

[11]Their paper also contains critical remarks on Yukawa's paper.

had developed a meson theory which was in some respects close to the Standard Model of particle physics, since it was factually based on internal gauge symmetry.

It later turned out (Conversi et al. 1947, p. 209) that the particle detected by Neddermayer and Anderson - to be called muon - could in fact not be the one predicted by Yukawa. But at that time meson theory of strong interactions had gained such a respectable status that it was soon proposed that besides the muons there also exist the Yukawa mesons. The confidence in the theory was soon rewarded: In the same year the Bristol group headed by Powell announced the detection of an additional meson (Lattes 1947, p.694).

Let us summarize the lesson from Yukawa's meson theory. This theory taken in a strictly logical context was not of outstanding value. Feynman (1965, p. 160)[12] describes the situation in the following words: "Yukawa guessed an idea for the nuclear forces in 1934, but nobody could compute the consequences because the mathematics was too difficult and so they could not compare his idea with experiment. The theories remained for a long time, until we discovered all these extra particles which were not contemplated by Yukawa, and therefore it is undoubtedly not as simple as the way Yukawa did it". But one may observe that the concepts by means of which Yukawa enriched the core-context proved to be essential for the further development of physics.

We end this case study by a curious extension of what has been called the Heisenberg-Einstein principle (Pandit 2002a, 2002b). Theory does not only tell us, what can be measured, but it often tells us, which experiment we consider relevant: Already in 1932 Paul Kunze detected the trace of a particle in a cloud chamber, which had too little ionizing power for a proton and too much for an electron (Kunze 1933). This implied that the mass should be between that of an electron and that of a proton. But the time was apparently not yet ripe to recognize the importance of this observation and the author remarked bluntly 'The nature of this particle is unknown'.

3.4   The $\theta - \tau$ puzzle

An instructive example of a puzzle, the solution of which created new problems and developments in the core-context *in* physics is the fa-

---

[12]Feynman, character p. 160

mous so-called $\theta - \tau$ puzzle. In the early 1950s two seemingly different kinds of particles were detected, which, however, had the same mass and – even more astonishing – the same lifetime. These particles were 'mesons': that is, medium heavy particles, much heavier than the electron, but lighter than the hydrogen atom. The reason why the two kinds of particles were assumed to be different was that they had different decay channels. One of them, the θ-particle, decayed into two lighter mesons (pi-mesons), the other, the τ-particle, decayed into three of them. This difference in their modes of decay implied that they behaved differently under space reflections.[13] Decay into two pi-mesons implied that the θ had positive internal parity, that is the associated quantum field of the θ did not change sign under space reflection:

$$\Phi_\theta(-x, -y, -z) = \Phi_\theta(x, y, z),$$

Whereas the decay of the τ into three pi-mesons implied that it had negative internal parity:

$$\Phi_\tau(-x, -y, -z) = -\Phi_\tau(x, y, z),$$

The puzzle to solve was: How does it come about that the particles which had different parity had the same mass and lifetime? The solution was rather unexpected and had far-reaching consequences. One did not find a mechanism which could explain why two different particles can have the same mass and the same lifetime. It was shown that (1) the θ and τ were not different but one and the same kind of particles and that (2) an important principle was violated, the principle which was a part of the core-context of all previous theories. From the apparent difference in parity of the decay products one can only then conclude a different parity of the decaying mesons, if the interaction which causes the decay respects symmetry under space reflections (conserves parity). The solution of the puzzle came by just giving up this seemingly obvious assumption inherited from the previous context.

The first decisive step towards the solution of the $\theta - \tau$ puzzle was made by T.D. Lee and C. N. Yang in 1956. They showed that all re-

---

[13]Under a space reflection the three space coordinates change sign: (x, y, z) goes to $(-x, -y, -z)$. Under such a space reflection the left hand becomes a right hand and *vice versa*. If "parity is conserved", that is symmetry under space reflection is respected, left and right handed objects are equivalent, if "parity is violated", that is the symmetry under space reflections is not respected, left and right handed objects are intrinsically different.

sults obtained so far for weak interactions, the interactions by which the θ and τ decay, remained unchanged if parity was not conserved. They proposed that non-conservation of parity was indeed the solution and they suggested experiments to test the hypothesis. Many physicists were sceptical wondering whether this would be the solution. When Mme C. S. Wu undertook to perform the experiments proposed by Lee and Yang, some made even jokes[14] about her doing complicated experiments where the outcome was predictably clear, namely that parity cannot be violated.

The experiments of Mme Wu (1957) and collaborators turned out, however, to show that parity was violated in weak interactions. And now the solution was universally accepted: Parity is violated and $\theta - \tau$ is indeed one and the same particle, no wonder that it has only one mass and one lifetime, even if the decay products have different parity. Before we come to the far-reaching consequences of this solution, let us shortly discuss the context in which alone the puzzle could occur and in which it had to be solved.

In 1896, H. Becquerel discovered radioactivity. It was soon realized that there are three kinds of radioactive radiation, called α, β, and γ rays. The β rays turned out to be electrons and to come from a decay of a nucleus into another one. In this decay an electron and an antineutrino, a (nearly) mass-less neutral particle, are emitted. These β-decays are essential in the processes of energy production in the sun. As mentioned in the previous case study, in 1934 Enrico Fermi developed a theory in the framework of relativistic quantum field theory which could explain the huge amount of data collected from those decays. The interaction responsible for β-decay, formalized in Fermi's theory, is called the weak interaction. It was generally assumed, but by no means proved, that the decays of the $\theta - \tau$ were also due to weak interaction. This assumption was only based on simplicity and order of magnitude arguments. One did not have a formal expression, not to speak of a numerical solution for the mechanism of these $\theta - \tau$ decays.[15] The proof, however, that parity is violated in weak interactions had to be given for those decays where a theory existed, namely the nuclear β-decays: the electrons emitted in those decays turned out to be predominantly left-handed, that is their spin turned in the

---

[14]J.H.D. Jensen (private communication).

[15]Only in the Standard Model developed in the 1970s, the detailed theory of these decays has been established.

opposite direction of their momentum. After this had been shown, it was immediately accepted that parity violation was the solution to the $\theta - \tau$ puzzle, though a quantitative description of the decays leading to the puzzle was missing.

We now turn to the far-reaching consequences and the problems which this solution created. It soon turned out that the parity violation inside weak interactions was not a small effect, but maximal. It fitted exactly in a very elegant, purely mathematically motivated, scheme developed by B. V. d. Waerden and H. Weyl. The formalism in this scheme is the so-called spinor representation of the symmetry group of special relativity, the so called Lorentz group. The essential objects in this formalism have definite handedness. This means that they violate parity by construction. This was the reason why Pauli (1933) in his famous handbook article had rejected the "Weyl-Spinors" as purely formal. He wrote: "These wave equations[16] are however not invariant under reflections (exchange of left and right) and therefore not applicable to physical reality. The absence of invariance of the wave equation manifests itself in a peculiar coupling of the direction of the spin-angular momentum and the current".

These very Weyl-Spinors are now basic in the Standard Model of particle physics and the solution of the puzzle has turned into a deep problem: *Why is parity conserved in some interactions?* Even deeper than this problem yet closely related to it is the following problem (see sect. 7.2.3): *What is the origin of mass?*

Let us summarize: The solution of the $\theta - \tau$ puzzle was not related to the puzzling particles θ and τ. On the contrary, it arose in the much larger context of weak interactions and symmetry properties of space. The solution turned out to create a problem in an even larger context, namely in the context of the Standard Model of particle physics.

We close this discussion with some remarks which show how the solution of the specific puzzle indeed changed the core-context of the theory. This has to do with symmetry under time reversal. Under time reversal the time coordinates change sign, that is t goes to $-t$. This implies that the past becomes future and *vice versa*. Though time reversal invariance is badly violated in our macroscopic world, all fundamental laws known up to then are invariant under time reversal. Therefore, it was generally accepted that though the symmetry under space reflections was violated, the symmetry under time reversal was

---

[16]This refers to the equations for Weyl-Spinors.

respected by all interactions and there were good reasons for that assumption.[17]

This symmetry under time reversal forbade for certain particles certain decay modes. When, however, such decay modes were observed by V. L. Fitch, J. V. Cronin and collaborators in 1964 (Christenson 1964), the scientific community had absorbed the consequences of the $\theta - \tau$ puzzle and nobody thought of introducing two different particles. But one assumed immediately that the symmetry under time reversal was violated. In contrast to the maximal violation of parity, the violation of the symmetry under time reversal is very weak and its origin is one of the fundamental unsolved problems of the Standard Model of particle physics.

---

[17]We assume the validity of the so called CPT theorem. It states that any interaction which is compatible with the postulates of relativistic quantum field theory is invariant under the combined application of space reflection (P), time reversal (T) and the transition from matter to antimatter. Consequences of this theorem have been tested with an extremely high precision.

Chapter 4

CORE-CONTEXT GUIDED THEORY DEVELOPMENT IN PHYSICS

[...] dass es der freie, in Symbolen schaffende Geist ist, der sich in der Physik ein objektives Gerüst baut, auf dass er die Mannigfaltigkeit der Phänomene ordnend bezieht. Er bedarf dazu keiner solchen von aussen gelieferten Mittel wie Raum, Zeit und Substanzpartikel; er nimmt alles aus sich selbst. H. Weyl 1949

4.1 Introduction

In this chapter, we shall bring out the importance of the dynamical development of the core-context of a theory in concrete problem-situations in physics. We shall especially investigate the disappearance and reappearance of important concepts in different core-contextual scenarios at the frontiers of theory development.

Our main laboratory of investigation in this chapter is the Standard Model of particle physics. Its salient features were set up in the 1960s and 1970s. Thus, on the one hand, this theory is recent enough not to be obscured by so many purely historical questions, but, on the other hand, it is mature enough to allow for a philosophical analysis, focusing on its dynamic complexities.

One of the most important principles in focus here is the principle of gauge invariance. It was introduced by Weyl (1918, 1919), one of the greatest and most influential mathematicians of the twentieth century. He did so in 1918 in the core-context of general relativity, but it proved to be important mainly in the core-context of quantum physics, especially relativistic quantum field theory. We do not intend here to give a historical description nor an account of the underlying physics. For that we refer to the excellent review by L. O' Raifeartaig and N. Straumann (2000) and modern textbooks of quantum field theory.

At present, a possible new core-context in the horizon is string theory with uncertain outcome. Intended to unify quantum theory and general theory of relativity, it is still a construction site and not a well-established theory. Nevertheless, its great relevance in the present study lies in the fact that it allows us to study again the disappearance and reappearance of important steps in theory development in different core-contexts. Interestingly enough, string theory was developed originally in particle physics in an attempt to replace quantum field theory in the sector of strong interactions, giving it a fundamental role. But from there it was finally thrown out as a fundamental concept and replaced by quantum chromodynamics, the quantum field theory of strong interactions.

As we have just pointed out, string theory is now an interesting construction site at the active frontiers of physics, where physicists are looking beyond the Standard Model of particle physics and trying to reconcile quantum theory and general theory of relativity. If string theory in its new appearance is successful, it will form an all-embracing core-context for particle physics and general relativity - a theory of everything (TOE). Quantum field theory will then be a limiting case of string theory.

Viewed within methodological structuralism, the present study analyses the inner development of physical theory and its characteristic methodological variances, proposing a model of the core-context guided theory-development and unification. Far ahead of the core-context building interfaces and irrespective of the state of the art of instrumentation and experimental possibility building up at the active frontiers of research, theories participating within this framework are expected to make new predictions and ask new probing research questions, beyond the reach of well-established existing theories.

4.2   From Newtonian Mechanics and Maxwell's Electromagnetic Theory to General Theory of Relativity

In order to illuminate the context of the main subject of our discussion, let us briefly sketch the core-contextual development of "classical" physics. With the overwhelming success of Newtonian physics, its principles were forming the classical core-context of theory development. Euler and Lagrange succeeded in incorporating the physics of continua – hydrodynamics for instance – in that core-context. However, a major challenge was to incorporate into it the seemingly unrelated electric and magnetic phenomena. This was attempted by establishing a mechanics of the ether, the supposed carrier of such phenomena. Seemingly these efforts were crowned with success by Maxwell's brilliant unified theory of electricity and magnetism. In the equations named after him, this theory took the final form. Although Maxwell had derived the displacement current, a cornerstone of his theory, employing ether-mechanical arguments, physics as a whole remained still divided between two quite unrelated core-contexts, that of mechanics, on the one hand, and electromagnetism, on the other. The presupposed ether seemed to have contradictory mechanical properties, as was shown by seemingly contradictory experiments. The observation of star aberration seemed to indicate the ether at rest with respect to the universe but in the famous 1987 experiments of Albert A. Michelson and Edward Williams Morley no motion of the earth through the ether was found.

Einstein accorded methodological priority to the more abstract field-theoretical core-context of Maxwell's equations, thereby giving up the principles pertaining to the core-context of classical mechanics, notably the ether as carrier of electromagnetic fields. In this way he was able to resolve the above mentioned contradictions. This lead to a modification of time honoured principles of mechanics. No wonder that it took some time to convince the physics community of the necessity of this field-theoretical core-contextual shift. In order to arrive at the major premise in Einstein's theory of (special) relativity, namely that all uniformly moving coordinate systems are equally well adapted to describe electrodynamics, the kinematics of Newtonian mechanics had to be modified substantially. But Newtonian theory of gravity stood irreconciled besides electrodynamics and the principle of relativity. There was a situation of two irreconciled - even contradictory – components in the classical core-context of physics: Newton's theory of gravity and Maxwell's theory of electrodynamics. Apart from minor discrepancies, such as the movement of the perihelion of the planet Mercury, which could possibly have many causes, there was no serious contradiction within Newton's theory itself. On the other hand, there were many unexplained facts in Maxwell's theory of electromagnetism. One example was the existence of the permanent magnets. In this context, the remarks Duhem made against Maxwell's theory (Duhem1906:132) are well justified. Generally, there were many unexplained effects which could only later be explained by quantum physics.

In the contradiction between the principle of relativity and the Newtonian theory of gravitation, Einstein sided again with the field theoretical concepts of Faraday and Maxwell. After he had thrown out the ether from the core context of physics, he went further and also gave up the time honoured and very successful concept of the action at-a-distance in gravity and created a new theory of gravity in his general theory of relativity. Special relativity, which replaced Newtonian mechanics by relativistic mechanics, where mass and energy became interchangeable, was based on the postulate that Maxwell's equations have the same form in all uniformly moving systems. The general theory of relativity postulates that all physical laws have the same form in all coordinate frames (for a more detailed discussion, see sect. 4.4.) In this way a theory of gravity was obtained which a *fortiori* was compatible with the more restricted principles of special relativity. The consequences were even more radical than those of special relativity: Not only absolute time and space were no longer valid concepts, but also absolute space with its Euclidean geometry lost its privileged role in physics. But there was still one serious problem left. There seemed to be in physics two equally important but unrelated cores-contexts: that of gravity on the one hand and electromagnetism on the other. There was no unification in sight in physics. At the frontier of physics, there were clouds over any further theory-development or theory unification.

4.3 Weyl's Attempt to Unify Electrodynamics and Gravity by 'Gauge Invariance'

In a series of path breaking papers (Weyl 1918: 384ff/1919:101)[1] Hermann Weyl proposed to obtain the synthesis of electromagnetism and gravity by extending the general relativity of motion through the relativity of size. In his book *"Space, Time and Matter"* he gives a concise outline of his reasoning (Weyl 1922: 299):

> With that we rise to a last synthesis. In order to characterize the physical state of the world at a point (Weltstelle) by means of numbers, one must not only relate the surroundings of this point to a coordinate system but also fix certain units of measure. It is necessary to obtain a statement equally on principle about this second point, the arbitrariness of units of measure (length), as made by the theory of Einstein with reference to the first point, the arbitrariness of coordinate systems. Just this idea ...... brought the final breakthrough to a pure infinitesimal geometry. We are forced to this further step by the relativity principle of size with equal necessity as to the first relativity principle of motion. ...... Besides coordinate invariance of the field equations entailing four arbitrary functions, there occurs as fifth invariance gauge invariance. ...... We may expect that the gauge invariance is related ...... to the 5th conservation law, the conservation of electric charge.

Although Einstein regarded Weyl's theory as mathematically brilliant, he argued that the principle of gauge invariance, as introduced by Weyl, was incompatible with the present state of knowledge concerning the spectra of atoms. In effect, the criticism amounted to identifying how much Weyl's principle of gauge invariance was out of step with the core-context of theory development in place at that time, since on this very principle the spectra of the atoms would have to depend on their history, leaving their stability and constancy unexplained. The discussion between Einstein and Weyl shaped itself

---

[1] Weyl, H. (1919, 101-133); earlier communications: Sitzungsber. d. Preuß. Akad. d. Wissensch.1918, 465, Math. Z. 2. For details see Straumann, N. (1987, pp. 414-421); O'Raifeartaigh, L. (1997); Ref. 9; see also Mielke, E. W. and Hehl, F. W. (1985).

at two levels, a phenomenological level where the argument proceeds mainly in terms of the constancy of atomic spectra (Einheituhren) and a more formal level where it builds in terms of the naturalness and beauty of gauge invariance. Thus, in a letter to Einstein, Weyl wrote: "If you are right, I am sorry to accuse God of a mathematical inconsistency". Picking up this argument, Einstein showed that Weyl's proposal of "*Nahgeometrie*" (infinitesimal geometry) could not be considered fully consistent, since it still preserved similarity (taken in a mathematical sense). With regard to the arguments concerning the atomic spectra, Weyl had to agree. Accordingly, he modified his theory in such a way that the ideal process of the transportation of a world line had nothing to do with the real behaviour of scales and clocks.

Pauli's (1921) remarks in his famous article on relativity are very interesting in this connection, since they too give a good description of the core-context of the general theory of relativity as compared to Weyl's extension. After describing Weyl's retreat to purely formal argumentation, Pauli wrote (Pauli 1921: 763):

> This renunciation seems very weighty. Even if now there is no direct contradiction to experience, nevertheless the theory seems from a physical point of view deprived of its inner powers of persuasion. For example, the relation between electromagnetism and world geometry is now basically not a physical one but a purely formal one. There is no longer an immediate relation between the electromagnetic phenomena and the behaviour of gauges and clocks. ... Incidentally, only formal, but no physical, reasons can be asserted for a relation between world metric and electricity, in contradistinction to the connection between world metric and gravitation, which finds its strong empirical support in the equality of heavy mass and inertia.

This could have easily been the end of a historical cul-de-sac. Weyl's theory could have ended in the dust bins of history. But this was not the case. On the contrary, there was a renewed interest in Weyl's theory. The main reason for this was that there were two branches developing from Weyl's original ideas on the principle of gauge invariance: One of them remained in the realm of classical physics but extended the number of space-time dimensions to five, thereby opening a new frontier of physics with extra-dimensions. This was initiated by the work of T. Kaluza (1918)[2] and O. Klein (1926) and regained new interest in physics through the advent of supergravity in the late 1970s. Later we shall return to this point for more discussion.

On the other hand, another development initiated directly by Weyl and seen by him as the true core of his theory finally turned out to be of paramount importance as a unifying principle in the Standard Model of particle physics. This particular phase of theory development due to Weyl had to wait until a new frontier beyond classical physics was inaugurated by quantum mechanics in the 1920s, bringing quantum field-theoretical core-context building and re-building to bear upon theory development. We shall later (in sect. 4.6.2) discuss how relevant the first of these branches, the introduction of extra-dimensions, is in connection with a possible core-context re-building in physics. Let us first consider the splendid triumph of Weyl's principle of gauge invariance in the quantum core-context. The methodological questions concerning theory-development in physics this is bound to raise will be briefly considered in Chapter 5.

4.4 Gauge Invariance in Classical and Quantum Electrodynamics

Already in 1922 Schrödinger (1922:13) saw significant connections between Weyl's "Weltgeometrie" (world geometry) and the (old) quantum mechanics of Bohr. Soon after the establishment of the Schrödinger equation (1926), Fock (1926: 226) and London (1927: 375) found that there is a remarkable invariance of the equations of quantum mechanics if electromagnetic interactions are included. Weyl took this up and postulated his gauge principle in the new quantum core-context of theory-development. In order to give a simple example, let us consider a state described by a Schrödinger wave function $\psi(x,t)$. The values of this function are in general complex numbers. The probability to detect a particle at the space-time point (x,t) is given by the absolute square $|\psi(x,t)|^2$. Hence the same physical situation is described by the wave functions ψ and ψ' if they differ by a factor of absolute value 1. Such a complex number of modulus 1 is called a phase factor.

---

[2]T. Kaluza, Sitzber. Preuss. Akad. Wiss. 1921, p.966. Kaluza was like Weyl a mathematician (and a linguist).

In electrodynamics it was known for some time that there is an analogous situation where several forms of not directly observable quantities lead to the same physical result: There are different potentials which nevertheless yield the same electromagnetic fields, i.e., which describe the same physical situation. This feature was very confusing at the time when electrodynamics was established. Nowadays also in electrodynamics such a transformation which changes the potentials but which leaves the fields invariant is called a gauge transformation. This name is appropriate, since the transformations depend indeed on space-time. Classical electrodynamics was known to be gauge invariant, since it could be formulated without taking recourse to potentials while sticking to fields. In quantum mechanics, however, there seemed to be a problem, since into the Schrödinger equation the potentials enter and not the fields. The same holds true for the Dirac equation, which replaces the Schrödinger equation in the relativistic case.

Fock (1926) and London (1927) had found that the basic equations of quantum mechanics stay invariant, if changes of the above mentioned phase factors in quantum states are related to the changes of the potentials in electrodynamics. Weyl immediately recognized that his "gauge principle" could be applied to that case, the relevant quantity being no longer the length scale, but the phase of the quantum mechanical states. In a nutshell, if one postulates that the phase factor in quantum mechanics can be chosen arbitrarily at each space-time point, there must exist an electromagnetic field, and the simplest (minimal) form of interaction is also fixed by this gauge principle. Thus, the symmetry principle here does not only fix the form of the interaction but establishes its existence, there being only one number open, and that is the strength of the interaction, to wit the electric charge. What is methodologically remarkable is this: In the quantum core-context of theory development, the principle of local gauge invariance turned out to be theoretically so fruitful, even if its first application in 1918, being obviously far ahead of its times, was found inadequate. Soon after the principle of quantization had been applied to mechanics, it was also applied to field theory (the so called second quantisation) by P. Jordan, O. Klein, W. Heisenberg, P. A. M. Dirac and others. Dirac and Heisenberg applied it to electrodynamics. Dirac had discovered an equation describing electrons in the electromagnetic field which incorporated the requirements of quantum mechanics as well as (special) relativity. Seen as a kind of unification,

Dirac's relativistic quantum mechanics (1928) resulted in quantum field-theoretical core-context re-building, since it itself demonstrates its resolving power in its new prediction that every charged particle must have an antiparticle. It should be noted, however, that Dirac's speculation on these antiparticles (holes in the Dirac sea) were first met with great scepticism by most of his colleagues.

Dirac's equation, incorporated in the context of a quantum field theory, successfully described the magnetic properties of the electron and put the theory of antiparticles on a firm basis. The first such particle, the counterpart of electron, came to be subsequently known as positron. It had been discovered experimentally in 1932 by C. D. Anderson (see Chapter 3, sect. 3.1.2). However, relativistic quantum field theory led to mathematical difficulties in the form of infinities which inhibited the further development of the theory until after World War II. S. Tomonaga, R. P. Feynman, J. H. Schwinger, F. J. Dyson and others found a way out to bypass, if not to solve, these difficulties (see See Schweber 1949). It was here that gauge invariance of the theory turned out to be absolutely necessary for fulfilling this task. This phase of theory-development culminated in the construction of the fundamental theory of electromagnetic interactions known as renormalised quantum electrodynamics.

As was noted above, the electromagnetic field is necessary in order to ensure the invariance under multiplication by phase factors, the latter forming the group called by mathematicians $U(1)$. Electrodynamics can aptly be called the gauge theory of $U(1)$. Gauge invariance became now definitely a part of the core context of quantum field theory, since it gave not only a rationale for the existence and form of electromagnetic interaction, but also proved to be essential for the aforementioned taming of the infinities by the procedure of renormalisation. Therefore, it was not astonishing that one tried to extend the gauge principles to more groups with a richer structure than the group $U(1)$.

4.5 Yang Mills "Non Abelian Gauge Theory" and the Standard Model of Particle Physics

In the year 1954, C.N. Yang and R. Mills (1954:191; 1954: 631) succeeded in extending local gauge symmetry from the phase factors $U(1)$ to the more complex (non-Abelian) group $SU(2)$. This was done on a purely classical level and the quantisation of the resulting field theory was unclear; it was only known that it was much more complicated than quantisation of the $U(1)$ gauge theory, that is electrodynamics. It was generally recognized that there was a need for action in particle physics. In quantum electrodynamics the agreement between theoretical predictions and experimental results reached a level never achieved before in exact sciences. But for the phenomena of strong interactions - e.g., the interactions which hold atomic nuclei together - no precise calculations were possible. In the late 1950s and early 1960s the situation of meson theory was very adequately described by Feynman (1965), see sect. 3.3. The predicted meson was found, but in addition to it many other ones, and "therefore the theory was not as simple as Yukawa had thought".

The existence of mesons was well established and qualitative considerations gave many valuable insights into the theory of strong interactions, but the discontent about the lack of precise quantitative results lead to the establishment of a (temporary) new frontier. Though the unification of weak, electromagnetic and strong interactions under the roof of quantum field theory was the driving force for the efforts of Yukawa, it was doubted that quantum field theory was useful for strong interactions at all. G. Chew, an important player in particle physics of the 50s and 60s, even anounced that quantum field theory in strong interactions will "fade away like an old soldier" (Chew 1961). The attempt to open a new core-context in particle physics was based on a series of papers by Heisenberg (1943, 1944, see also Wheeler 1937) on what was called S- matrix theory.[3] They were written before the infinities in quantum field theory (see sect. 3.3 and sect. 4.4.) were tamed. Heisenberg thought that quantum field theory had to be modified essentially and therefore, before the new theory was found, he proposed to make use of all principles which would survive in the new improved quantum field theory. In the program of Chew and his collaborators this temporary program was taken as the final goal of particle physics. In this program one gave up looking for a field theory of strongly interacting particles and tried to treat all these particles like protons, neutrons, and mesons on the same footing. Due to this aspect the name *nuclear democracy* was coined. For quantitative calculations a method based on self-consistent solutions respecting general features of the S-matrix was developed. These general features were the sym-

---

[3]The S comes from Streuung (scattering)

metries of special relativity, internal symmetries like isospin (see sect. 4.3.) and general principles of quantum physics, like the superposition principle and the conservation of probability. Since the properties of the observed particles should be fixed by self-consistency, this approach was called *bootstrap* approach (to tear oneself up by one's own bootstraps).

Most important tools in nuclear democracy were the symmetry considerations, since they allowed the physicists to classify the known elementary particles and even predict the existence of new ones. Since symmetry considerations belonged to the few methods allowing some quantitative predictions, it was quite natural to try to apply the gauge principle to these symmetries. Indeed, already in 1938 O. Klein (1938) proposed a theory of strong (and weak) interactions which in the newer core-context would be called (broken) gauge symmetry of $SU(2)$. Klein's paper also showed a close connection between the gauge principle and the extension to higher dimensional spaces.

As regards the application of local gauge theory to weak interactions, in 1934 Fermi had developed a theory (see sect. 3.3). It was reformulated and especially adapted to the observed phenomenon of maximal parity violation by the establishment of the so-called V-A theory. This theory was phenomenologically very successful and it was by no means any 'falsification' but only core-contextual, theoretical, reasons which led to the attempts to replace this theory. Since the 1930s there was a somewhat diffuse discontent with the theory since the interaction was a contact interaction and not mediated by a quantum field. Later on, after the establishment of the renormalization procedure in quantum electrodynamics, there emerged a second, more precise reason for methodological dissatisfaction with the theory: The theory was not renormalisable. Many attempts were made to overcome the theoretical inconsistencies, some of them pursuing the path shown by local gauge symmetry. In order to show how deeply rooted local gauge symmetry was in the core-context already in place then, at least for a part of the physics community, we shall shortly describe how serious was the obstacle to applying gauge theory, but which nevertheless did not deter its advocates to try to save it, for a more detailed description see sect. 7.1.2 and 7.1.3.

The principal phenomenological difficulty was that in a local gauge theory the field quantum which ensures gauge invariance and which mediates the interaction is massless. This strict result was of course very welcome in the case of quantum electrodynamics where the field quantum, the photon, is indeed massless. This property is essential for the taming of the infinities in quantum electrodynamics, which in turn is necessary in order to arrive at precise predictions. However, neither in strong nor in weak interactions were any such massless particles observed. In strong interactions the masses of the field quanta (mesons) fitted neatly into the spectrum of the other nuclear particles. In weak interactions the data indicated that they should – if they existed at all - be much heavier than all elementary particles known at the time.

On the formal level, the quantisation of Yang-Mills type theories was finally achieved in the early 1970s by G. t'Hooft and M. Veltmann. The obstacle of mass-less gauge fields was overcome in strong and weak interactions in two different ways: In strong interactions, Fritzsch, Gell-Mann and Leutwyler (1973: 365) proposed "that there are several advantages in abstracting properties of hadrons and their currents from a Yang-Mills gauge model based on coloured quarks and a colour octet gluon". The fundamental fields of the theory are quarks and gluons, the latter being the field quanta of the gauge field of strong interaction. They are also massless. As the term "abstracting properties" indicates, quarks and gluons cannot be observed directly, they are "confined" forever in the observable particles, e.g., protons and neutrons. The gauged symmetry group, $SU(3)$, is called colour-group, since colour was the whimsical name given to the new property related to this group. The theory demonstrates its high resolving power by raising new questions and making powerful predictions. Some examples are the explanation of the so-called *asymptotic freedom* and deviations from it, the prediction of new types of events (*three-jet events*) and the success in classification and mass determination of new particles (*heavy mesons, quarkonia*). We shall comment on that in more detail in Chapter 7.

Though this new approach meant a return of quantum field theory to the theory of strong interactions, nevertheless some principles of nuclear democracy, which was discussed above, remained and were essential for the development of the new theory. This is explicitly acknowledged by the autors of the above mentioned seminal paper (Fritzsch, Gell-Mann and Leutwyler 1973). They write: "The resulting picture could be equivalent to that emerging from the bootstrap-duality approach". In this paper the field theoretical aspect is not stressed particularly and the proof of renormalizability of such a non-

Abelian gauge theory is not even mentioned. This is presumably due to the prevalence of the core-context of "nuclear democracy" over that of quantum field theory at that time.

In weak interaction the problem of massive gauge bosons was only overcome after another concept entered the core-context of particle physics, namely the concept of spontaneous symmetry breaking by the introduction of a new particle, the so called Higgs boson. The limitations of space do not allow us here to go into full details. It should suffice to say that in such a symmetry breaking the fundamental equations respect the symmetry, but not the ground state of the theory, the so-called vacuum state.[4]

In the field of weak interactions, already in 1967, before the renormalizability was proved, S. Weinberg (1967) posed the question: "What could be more natural than to unite these spin-one bosons, that is the particles mediating the electromagnetic and the weak interaction, into a multiplet of gauge fields?" He enumerates the problems, alluded to above, and concludes that the problems might be avoided by introducing the photon and the intermediate fields of weak interactions as gauge fields of one gauge group, in the hope that the model may be renormalizable. The final theory of weak interactions is a spontaneously broken local gauge theory based on the group $SU(2) \otimes U(1)$. By virtue of its high resolving power, the theory made finally remarkable predictions concerning the (i) existence of neutral currents; (ii) existence of very heavy intermediate Bosons; (iii) existence and mass of the top quark; and (iv) the existence of a Higgs boson.

All these predictions have been confirmed. As a consequence, all other theories of weak interactions just faded away. Strong, weak and electromagnetic interactions are thus all described by gauge fields. The local gauge transformations are at the core of what is known as the Standard Model of elementary particle physics. It is a local gauge theory of $SU(3) \otimes SU(2) \otimes U(1)$, the first factor $SU(3)$ pertaining to the strong and $SU(2) \otimes U(1)$ to electroweak interactions. Thus, the developmental trajectory of Weyl's gauge invariance principle reaches out to the very core of the Standard Model where all interactions are consequences of and subject to gauge symmetries. The importance of gauge theories for physics stems from the tremendous success of the mathematical formalism in providing a unifying framework for

---

[4]For a discussion of the problems of spontaneous breaking of a gauge symmetry see e.g. Dosch et al. 2011

describing the fundamental interactions of electromagnetism, the weak force and the strong force in the language of local gauge theory. It makes numerous predictions, many of them having been confirmed experimentally. Several minor discrepancies have been observed but some of these have disappeared. Many physicists are disappointed over the fact that the Standard Model is too good, since it seems to leave little room for new physics.

4.6 Beyond the Standard Model of Partcile Physics: Extension or Creation of a New Core-Context?

As a core-context of theory development with very high resolving power, a physical theory may not look so impressive, particularly to those who only look for solutions instead of new probing questions with the potential of opening up a new frontier of discovery. The real situation can be better understood and appreciated in terms of the concept of resolving power. The high resolving power of the Standard model of particle physics, along with completely new research questions it itself poses and their possible solutions, will form the core-context of theory development in the future. We list here five of the most decisive areas of problem and theory development:

(1) The gauge group of the theory is a direct product of three groups, leading to 3 different couplings; thus there is no true unification of the three interactions: strong, electromagnetic and weak.

(2) There are three groups of quarks and leptons, called families. The theory does not explain why there are so many fundamental fields and just three families.

(3) Why is nature dominated by gauge interactions? Though the gauge concept originated in the strict infinitesimal geometry ("Nahgeometrie") of Weyl, the gauge principle somehow goes beyond the concept of strict locality.

(4) Why does fermionic matter exist at all? Also the origin of mass remains an unsolved problem and this is the most concrete and most challenging problem.

(5) The theory is not compatible with the general theory of relativity. Incorporation of gravity into the theory would make it un-renormalisable. Closely related to this is the problem of the cosmological constant (see sects. 4.6.1, 4.6.2.).

Problems (1 - 3) can perhaps be solved in the not too distant future, at least to some extent, by staying in the framework of the present core-context while just extending it. The last problem (5) brings two core-context-building concepts, namely local quantum field theory and general relativity, together yet into conflict. The former is the core of the Standard Model of particle physics, the latter the core of the Standard Model of cosmology. It should be clear that a new core-context, developed with a view to incorporating the two precursor core-contexts, should contain them as specializations under certain conditions. We stress that we have only addressed the new problems which a theory's resolving power brings to the attention of the physicist, but not those which emerge with new experimental findings, e.g. the neutrino mass. There is, of course, the hope that the latter may help to solve the former ones.

4.6.1 Extending the Core-Context of Theory Development

In order to extend the Standard Model of particle physics the most promising direction at the moment involves extending the gauge group. In the present Standard Model of particle physics the underlying gauge symmetry is a product of three groups: A group $SU(3)$ for the strong interactions, the so-called chromodynamics, an $SU(2)$ for the weak and an $U(1)$ for the electromagnetic interaction (this is oversimplified, because weak and electromagnetic interactions do mix). This entails that there is a separate coupling constant for each interaction. The quest for unification, a driving power for the dynamical core-context development thus drives the theory to find a single group with only one coupling constant for all three interactions.

This might seem a futile hope, given the large differences between the couplings, already indicated by the names (weak and strong). The hope for a fruitful extension of the present Standard Model is reaslistic due to a peculiar feature of it, namely the asymptotic freedom. Asymptotic freedom is a property of non-Abelian gauge thories (see sect. 7.2): With increasing energy the coupling strength diminishes.

D. Gross, who with F. Wilcek was awarded the Nobel prize for the discovery of asymptotic freedom writes (Gross 2005):

> "the apparently insurmountable barrier to unification - namely the large difference in the strength of the strong interactions and the electro-weak interactions - was seen to be a low energy phenomenon. Since the strong interactions decrease in strength with increasing energy these forces could have a common origin at very high energy. H. Georgi, H. Quinn and S. Weinberg showed that the couplings run in such a way as to merge somewhere around 10^{10} Gev. This is our most direct clue as to where the next threshold of fundamental physics lies, and hints that at this immense energy all the forces of nature, including gravity, are unified".

This provides another example showing the resolving power of the Standard Model of particle physics.

Remember that at the time when Yukawa introduced the field theory of strong interactions (see sect. 3.1.2.) the idea to model strong interactions according to Fermi's theory of weak interactions was a guiding principle. In that case, however, progress was only possible because Yukawa gave up this idea, except for the unifying methodological frame of quantum field theory. This again is a nice example showing the complexity of the dynamical core-context development.

For some time one had hoped that $SU(5)$ would be the final unifying gauged symmetry group. But in spite of some beautiful features this possibility is practically ruled out as it stands now. The reason is that it predicted the decay of the proton with a lifetime which is now experimentally excluded. But there is such a great confidence in the new core-context that one does not give up hope of unification. One still makes huge experiments in order to detect proton decay, even if there is by now not the slightest experimental hint for such decay. Some of these huge experiments yielded, however, another very important result, leading to core-context development at another frontier. These experiments showed that different kinds of neutrinos transform into each other (*neutrino oscillations*) which in turn indicates, that at least one neutrino has a (tiny) mass and that the Standard Model of particle physics has to be modified.

The most promising candidate for the extension of the Standard Model is now a unified supersymmetric gauge symmetry. Supersymmetry is an extension of the symmetry concept which assumes invariance under the transformations which connect particles of integer spin (bosons) and half integer spin (fermions). There is a theorem (Coleman and Mandula 1967, 1251) that such combination of external and internal properties is impossible with normal symmetry groups. But the supersymmetric extension turns out to be of the kind which finds a loophole to escape this restriction (Haag et al. 1975, 257). The 'supersymmetric standard model' is in accordance with all observed phenomena. However, it also makes specific predictions which should be tested in the near future, say within the next ten years. The predictions include a definite lifetime for the proton and the existence of a new kind of particles, the so-called supersymmetric partners of known particles. If the proton decay is observed, this will be generally hailed as the "proof" of the "grand unification" , and an observed supersymmetric partner as a proof for supersymmetry. Of course, here physicists are far from being so naïve as to take "proof" in the standard mathematical sense. But what is implied is that only the supersymmetric grand unified Standard Model could form the new core-context of particle physics. If this new core-context does not develop further resolving power, the physicists might lose interest in it.[5]

A similar situation occurred with the general theory of relativity during the 1940s. Though it was highly admired, it was certainly not at the centre of interest, since it apparently developed no resolving power commensurate with the theoretical and experimental possibilities of the time. However, it did develop its resolving power only during the second half of the 20th century with further developments in theory-experiment interface building.

4.6.2 Bringing a New Core-Context in?

There is a far more radical approach to changing the Standard Model of particle physics which is known by the name of 'string theory'. String theory is not a theory proper with an axiomatic formulation but rather a construction site of theory development. In a meta-theoretical approach, it could hardly attract the attention of the philosopher. Nevertheless, it forms at the moment a very active part of fundamental physics and aims at developing the core-context of microphysics of the 21st century. In what follows, far from even attempting a superficial

---

[5]Up to now (July 2012) there are no convincing indications for "New Physics"

overview, we shall only highlight some aspects of it in the context of our foregoing discussion. It is interesting since it shows - if it is finally successful - some parallels to the development of the gauge principle.

The origin of string theory is traceable to a development of particle physics which was very much *en vogue* in the 1960s, namely 'nuclear democracy' and bootstrap-duality (see sect. 4.5.). In this approach, one wanted to give up the concept of local quantum field theory for the treatment of strong interactions. The proliferation of the candidates for the status of "elementary particles" was treated by proclaiming that all were equally fundamental, implying that there are no "elementary" particles at all. In 1968, G. Veneziano invented a model, which represented essential features of this approach in a simple formula (Veneziano 1968, 190). In the further development of the Veneziano model it turned out that it corresponded to a theory of strings, that is two dimensional extended objects. Initially there was much excitement about string theory as a theory of strong interactions, but it was short-lived. There were several reasons for this: There were some open contradictions with experiment, and it was not possible to make precise quantitative predictions. The most decisive reason for a rather abrupt end of the interest was however the rise of Quantum Chromodynamics (QCD). It was around 1973 that QCD gained recognition as the correct quantum field theory of the strong interactions, throwing out the string theory.

Among the reasons which made string theory inappropriate as a theory of hadrons the following are noteworthy: (i) the large number of extra space-time dimensions demanded by string theory in order to be consistent; and (ii) the existence of mass-less particles with spin 1 and spin 2 in the spectrum of string states. Nevertheless, string theory has survived, now as a theory of everything (TOE),[6] and no more as a theory of strong interactions. One of the reasons for its survival is its inherent mathematical beauty.[7] Some of its deficiencies as pure strong interaction theory have turned out to be a blessing in disguise within a more comprehensive scheme, so the occurrence of the mass-less spin 2 particle implies the incorporation of gravity.

Unlike quantum field theory, string theory is not based on locality.

---

[6]For general literature on string theory, additional dimensions and Standard Model of particle physics, see Green, Brian (2000), Randall, Lisa, (2005) and Veltman, Martinus (2003).

[7]One of the principal actors in the field, E. Witten, was awarded the highest distinction in mathematics, the Field medal.

Yet it has developed as a radical generalization of quantum field theory towards the end of 1960s and through 1970s and 1980s onwards. The fundamental objects in the theory are extended, one-dimensional open or closed lines (strings). In the old string theory the extension of the strings was assumed to be of nuclear dimensions, i.e. ca. 1 femto meter (10^{-15}m). In the new string theory, the theory of everything, the string extension is of the so-called Planck length, which is about 20 orders of magnitude smaller (10^{-35}m). At this distance the quantum-gravity effects are supposed to become significant. The ratio of the new to the old string length is much smaller than the ratio of the diameter of an atom to that of the sun. Therefore, the excitations of the strings, which would form a new kind of particle are of such high mass that we cannot imagine how to detect them even in the distant future.

There are, however, features of string theory, besides its interesting mathematical structure, which may provide a possible candidate for a future core-context: In the limit of string length going to zero, the result is a gauge theory containing (spin 1) gauge bosons as already present in the Standard Model of particle physics as well as the (spin 2) gauge bosons of the general theory of relativity. String theory is, therefore, a strong candidate for a unified theory of the Standard Model of particle physics and that of cosmology. As a consequence, it might strongly modify the current speculations regarding the origin of our universe, the so-called 'Big Bang'. In the frame of string theory even "pre-Big-Bang" scenarios have been developed. For decades ahead, it should be an active frontier in physics where a new core-context of theory-development might take over.

Chapter 5

SOME CONSEQUENCES FOR FOUNDATIONAL RESEARCH AND METHODOLOGY

Wenn ihr von den theoretischen Physikern etwas lernen wollt über die von ihnen benutzten Methoden, so schlage ich euch vor, am Grundsatz festzuhalten: Höret nicht auf ihre Worte, sondern haltet euch an ihre Taten! Albert Einstein (1934:113).

5.1 The Principle of Gauge Symmetry

Having described in Chapter 4 some of the developments at the frontiers of physics, the stage is now set to draw the consequences for the foundations and methodology of physics. It becomes now quite clear in terms of our dynamic core-context building model (turn to the formalizing model proposed in Fig. 5.1) that there is a strong undercurrent of core-contextual continuity in physics where many (Thomas Kuhn 1962, 1970a, 1970b, 1974, 1977, 1981, 1983, 1989a, 1989b, 1990a, 1990b, 1991a, 1991b, 1992, 1993, being a good example) tend to see punctuating discontinuity in the progress which physical theory makes in the course of theory change by theory development. One of the most important formal principles in the continuous development is the gauge principle[1], originally introduced by Weyl in 1918 in an attempt to unify electrodynamics and gravity. Being of great significance to fundamental physical theories, as the principle of gauge symmetry is to quantum field theory, any methodologically sound and dynamically adequate account of theory-development and unification in physics of the past 100 years cannot ignore this principle. Nor can it ignore the

---

[1]Even in numerical calculations for the Standard Model of Particle Physics the gauge invariance turned out to be essential (Lattice gauge Theories)

117

attempt of Kaluza and Klein to extend the general theory of relativity to five dimensions. Even though initially unsuccessful, these attempts eventually played a crucial role in the further evolution of physical theory, at least in the more speculative branches .

Gauge symmetry has become a unifying principle in the dynamic quantum core-context building, finally emerging as one cornerstone of the Standard Model of particle physics. In this evolution, it was Weyl's own subsequent work, published in 1929, which played a decisive role in bringing the theory of gauge invariance back into play in the fundamental physical theory.

5.1.1 Another Look at Theory-Development and Unification

In the preceding Chapters we have considered the important concept of gauge invariance under the methodology of dynamic core-contexts of theory development. What kind of methodological consequences can we now draw from our discussions? Both physicists and methodologists can, no doubt, draw valuable methodological consequences from the failed hypotheses, theories and experiments wherever in the historical development of physical theory they have left their footprints. This is particularly true where theory-unification is the main motivation behind theory-development even if it fails to take off at a particular point in time for no other reason than the lack of a proper *core-context of development.*

Modern theory-unification started with Newton's unification of terrestrial and cosmological mechanics. We have noted that another significant continuity in theory-development is discernible from Maxwell's (1865) theory of electromagnetism to the contemporary relativistic quantum field theory, which is the framework of the Standard Model of particle physics. Beyond this, significantly enough, the speculative superstring theory is a construction site at the active frontiers of physics. Thus, they provide both a complex and a rich scenario of core-context guided development of physical theory, demonstrating how in physics method and theory, foundations and the frontier, develop together in an intertwined manner. As a consequence, it is not possible to ignore two things as regards theory-development in physics. On the one hand, there always build up within it strong correlations between the dynamic core-context and the frontier of theory development, as also between the aim and structure of physical theory. On the other hand, it is these correlations which serve, in final analysis, as a testing-ground of our model of theory-development and unification. No surprise, if the dominant styles of thinking with regard to method, particularly those which are fashionable among philosophers of science, turn out to be highly one-sided. They fail to reflect these very correlations. This is generally true of all of them, even the critical rationalist Karl Popper (1983: 131) and other leading philosophers.

Since physical theory itself develops by core-context re-building, possibly marked by successful unifications, it becomes necessary to analyse the past attempts at unifications by relating them to their correlated core-contexts, thereby deepening our understanding of theory-development and unification within the methodology of core-context guided theory-development. The core-context is essential for the development and the acceptance of a theory, particularly where it plays an important role in opening up new frontiers, even in understanding the past developments whose philosophical significance was not clear before. If a theory is in the current core-context, any counter-examples to it are at first likely to be considered as indications for small modifications (or even as indications that something has to be discarded, as Einstein did with the experiments by Kaufmann, "disproving" special relativity), whereas a theory outside the core-context is easily called disproved, given a single counter-example. Within the scientific community, corroborations of "good theories" are generally taken as kinds of "proof" when considered in the appropriate core-context. A more recent example is the observation of three-jet events, very often cited as "proof" of the existence of the gluon. As a matter of methodological policy, an experiment and its results must be seen in the full core-context of the theory and theory-experiment interface which is connected with its framework.

Throughout our discussion, we have assumed, clearly not without justification, that the inner development of Standard Model of particle physics can be better understood from the perspective of methodological structuralism. We think that dynamic core-contexts are already at work in theory development and unification in particle physics. Think of the role quantum field theory plays in the Standard Model of particle physics. Similarly, the kind of fundamental role general theory of relativity plays in astronomy and the Standard Model of cosmology is not only remarkable but methodologically highly instructive.

In a sense, the present situation with regard to the Standard Model of particle physics has certain remarkable features which are similar to those we find in the case of general theory of relativity on the one hand and the classical field theory of Maxwell nearly 100 years ago on the other. Yet it is definitely far more complex and methodologically richer in so far as we have here in the Standard Model of particle physics a core-context building interface with cosmology in studying the early universe. The Standard Model of particle physics emerges as a good example of dynamic core-context re-building where not only the fundamental framework and frontier of physical theory co-evolve but theory-development and unification go hand-in-hand. In a somewhat over-simplified schematic form, this can be demonstrated with the help of the following block diagram Fig. 5.1

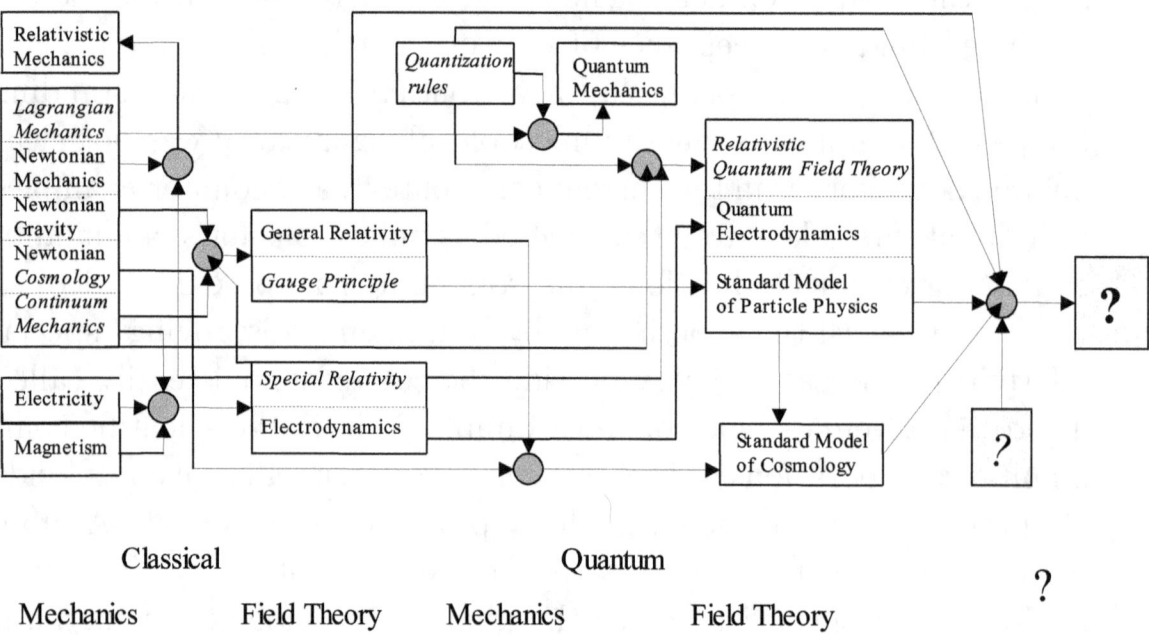

Figure 5.1: A block diagram giving a schematic view of core-context building across physics over the last centuries

Going back to classical mechanics and classical field theory, the block diagram schematically disentangles the core-context building interfaces which are characteristic of theory-development and unification at the frontiers of fundamental physical theories[2]:

(I) from classical field theory (continuum mechanics) and electricity and magnetism to electrodynamics, including special relativity;

---

[2]See Dosch, H. G. (2005a; 2005b).

(II) from Newtonian mechanics and special relativity to general relativity;

(III) from classical mechanics, in the form of ' higher mechanics' developed by Lagrange, Hamilton and Jakobi with the rules of Planck, Bohr and Einstein to quantum mechanics;

(IV) from classical field theory, in the form developed by Euler and Lagrange, and special relativity together with quantum physics to relativistic quantum field theory;

(V) from relativistic quantum field theory and electrodynamics to Quantum electrodynamics;

(VI) from relativistic quantum field theory with the gauge principle to the standard model of particle physics.

The Fig. 5.1 shows only the theoretical flow of scientific developments as if it was possible to leave out theory-experiment interfaces from our account. Of course, all these developments and the consolidation inside the fields are essentially determined by theory-experiment interfaces and experimental observations. As the point has been underlined in our discussion in Chapter 4, it is still an open question how the theory of general relativity and quantum physics can be united. Closely related to this problem is the unification of the Standard Model of cosmology with that of particle physics. Since it is not yet solved, the question formulating this problem cannot be answered in the metalanguage of the theory; strictly speaking it cannot even be asked, since there are two completely different languages at work, namely that of quantum field theory and general relativity. It is not even clear if the programme of "quantization of gravity", as it is normally called, is adequate. It might well be that both the principles of quantization and those of general relativity have to be modified essentially. Nevertheless, to disregard this problem as if it was lying outside the realm of philosophy of science would be an incredible impoverishment of the latter. The question of unifying quantum physics and general relativity concerns both the fields of particle physics and cosmology. It again raises a most important question of core-context of theory development, with far-reaching implications for the foundations and methodology of physics.

It is the dynamical core-context of theory development which is very important not only for problem-determination but for the assessment and even interpretation of problems in science. A very good example is to be found in the problem of vacuum energy. In the beginning of quantum field theory, this problem was considered as a major problem, especially because it turned out to be infinite. With the development of renormalized perturbation theory of quantum field theory it was, strictly speaking, no longer a problem, since it had no observable consequences. Nevertheless, many physicists felt that the problems of vacuum self energy were not really solved, but only swept under the rug. Indeed, it started again to play an essential role when the connection between quantum field theory and gravity came into focus, problematizing the very foundations of physics. The vacuum self energy plays an important role in the theory of gravity as so-called "cosmological constant". Therefore, it again plays a central role in quantum field theory.

We have treated in some detail Weyl's (1918) attempts to unify gravity and electromagnetism by generalizing Riemannian geometry (sect. 4.3.). Weyl conjectured that *Eichinvarianz* (invariance under the change of scale or "gauge") might in addition also be a local symmetry of the general theory of relativity. Concerned as he was with the unification of geometry and physics themselves such that the structure of matter and cosmos could be explained by means of a single universal world law, was not Weyl really anticipating a theory of everything, assuming of course as if there were no forces other than those known at that time? There is a hint to this effect in the following statements of Weyl (1918a: 170; Preface dated "Easter 1918"; Weyl, 1918b: 385, footnote 4):

> Again physics, now the physics of fields, is on the way to reduce the whole of natural phenomena to one single law of nature, a goal to which physics already once seemed close when the mechanics of mass-points based on Newton's *Principia* did triumph. Yet also today, the circumstances are such that our trees do not grow into the sky. I am bold enough to believe that the whole of physical phenomena may be derived from one single universal world-law of greatest mathematical simplicity.

This gives the impression as if Weyl's (1918, 1919) attempt had assumed a certain conception of physical theory within a rather highly speculative theory of everything. However, while admiring Weyl's mathematical genius, Einstein and Pauli criticized his theory for its unphysical results (sect. 4.3). From a superficial look, it seems that Weyl's early theory-unification failed completely because it did not fit the experiments. As we have noted in our discussion above, there are many examples where a theory failed to take off because it was developed in the wrong core-context. Thus, at a deeper level, there was a lurking tension between the core-context on the horizon then, the general theory of relativity, and Weyl's assumption of generalized gauge invariance. The dynamic core-context, where it is well-articulated, always includes theory-experiment interfaces, experimentally known facts and instrumentation, besides the general theoretical framework guiding and shaping the structure and development of physical theory. The situation, when over-simplified, can also be understood as one where negative feedback consequences warrant a theory's rejection on physical, foundational and methodological grounds. More modestly stated, Weyl's (1918, 1919) attempt can be understood as an attempt to extend the new dynamic core-context on the horizon then, viz., the general theory of relativity. Extending the idea of local symmetries, he tried to unite with it electromagnetism freed from the mechanics of Newton. As we have noted above, not only did it fail to gain acceptance, evidently it did not fit into the core-context on the horizon consistently. However, even in his first attempt at theory-unification, we can say, Weyl had philosophically and methodologically the best of the insight into theory-development by core-context re-building. Ideally, at a particular time, in physics one could not afford to be guided by the seemingly two unrelated core-contexts.[3]

Yet if a single instance of inconsistency, as cited by Einstein, was enough to set in motion a process of criticism and rejection, preventing Weyl from realising his 1918 programme of theory-unification, what is really of far-reaching methodological significance is the following fact: With quantum core-contextual development, Weyl's theory of gauge-invariance emerged as a principle of theory-development in its own right, making a unique contribution to quantum field theory, which is the methodological framework of the Standard Model of elementary particle physics. As Weyl wrote later on in a letter to Carl Seelig (quoted in Straumann, N. 1987, p. 421)

---

[3]It was, therefore, quite reasonable on Weyl's part to take the general theory of relativity as the newest core-context if one wanted guidance for theory development and unification.

Dating back to 1918, I undertook my first attempt to develop a unified theory of gravitation and electromagnetism, namely the theory based on the principle of gauge invariance which I put alongside the theory of coordinate invariance. Long ago, after quantum theory retrieved gauge invariance, the theory's correct core, as a principle which does not relate gravitation but the wave field of the electron with electromagnetism, I have myself given up *that* theory. (Emphasis by the author).

The development of "new" quantum mechanics towards the close of 1920s, replacing the rules of Planck's and Bohr's old quantum mechanics, re-activated the newly opened frontier of research in atomic, nuclear and elementary particle physics, all of them being areas where the possible quantum effects of gravity are too weak to be detected experimentally. It also triggered an intense foundational debate on physical theory and physical reality among the leading physicists themselves (Einstein, Podolsky and Rosen 1935; and Bell 1987).[4] There lay at the very core of Weyl's original proposal the concept of gauge invariance. Recaptured in the quantum core-context, it not only survived but proved to be indispensable for the further inner development of physical theory. As Weyl recalls:[5]

---

[4]See A. Einstein, B. Podolsky, and N. Rosen (1935, 777-780). The paper discusses pairs of quantum systems in interaction, showing how they remain entangled, or correlated, once they have separated and there is no more a possibility of any physical interaction between them; and J. S. Bell, (1988, p. 174). It is not surprising that until 1960s research on gravitational interactions remained largely outside this frontier. However, thanks to quantum core-context building and re-building, the focus of research remained on the unification of the other interactions: (i) the strong interactions among nucleons which build the various nuclei; (ii) the electromagnetic interactions among charged particles which form the atoms and molecules; and (iii) the weak interactions, governing the β-decay of the neutron and different radioactive elements, which play an important role in the evolution of stars.

[5]English translation of the original in German (in Carl Seelig 1960, S. 274) as quoted in Straumann, N. (1987, S. 421), "Einstein war von Anfang dagegen, und das gab zu mancher Diskussion Anlass. Seinen konkreten Einwänden glaubte ich begegnen zu können. Schließlich sagte er dann: "Na, Weyl, lassen wir das! So - das heisst auf so spekulative Weise, ohne ein leitendes, anschauliches physikalisches Prinzip - macht man keine Physik!" Heute haben wir in dieser Hinsicht unsere Standpunkte wohl vertauscht. Einstein glaubt, dass auf diesem Gebiet die Kluft zwischen Idee und Erfahrung so groß ist, das nur der Weg der mathematischen Spekulation, deren Konsequenzen natürlich entwickelt und mit den Tatsachen konfrontiert werden müssen, Aussicht auf Erfolg hat, während mein Vertrauen in die reine Spekulation

From the beginning Einstein was against it; and that gave rise to quite a few discussions. I believed in my ability to counter his direct objections. Finally, he said: Well, Weyl, let us stop it! In this way, - that is in such a speculative manner, without a guiding intuitive principle - one does not make physics. Today we have exchanged our points of view in this respect. Einstein believes that in this field the gap between idea and experience is so large that only the way of mathematical speculations has a chance of success; of course, consequences of the latter have to be elaborated and to be compared with experience. Whereas my confidence in pure speculation has sunk and a close connection to quantum-physical experience appears to become imperative; more so, since in my opinion it is not enough to unify gravitation and electromagnetism. The wave fields of the electrons and all that which may exist as irreducible elementary particles must be included.

Weyl's attempt (1918) was not only admirable and methodologically very bold but intimately related to the newly emerging core-context of Einstein's general theory of relativity which gradually changed Einstein's own earlier methodological thinking on the foundations of physics. The general theory of relativity, a theory of very high resolving power and, therefore, highly promising in terms of new predictions, had to wait for a long time for an appropriate theory-experiment interface to develop so that the consequences of the theory could be developed for experimental testing. However, this did not come in the way of Einstein himself pursuing the goal of a unified field theory until the very end of his career, though all in vain. What is striking in the evolution of their methodologies from 1920s onwards is how Einstein and Weyl kept on exchanging their roles: Einstein old was advocating a speculative approach quite contrary to the non-speculative approach of Einstein young and Weyl old; and Weyl young was advocating a speculative approach quite against the

---

gesunken ist und mir ein engerer Anschluss an die quanten-physikalischen Erfahrungen geboten scheint, zumal es nach meiner Ansicht nicht genug ist, Gravitation und elektromagnetismus zu einer Einheit zu verschmelzen. Die Wellenfelder des Elektrons und was es sonst noch an unreduzierbaren Elementarteilchen geben mag, müssen mit eingeschlossen werden".

non-speculative approach of Weyl old and Enstein young.[6]

Weyl's 1918 "unified theory of gravitation and electromagnetism" may be viewed as a grand programme to set up a new core-context of theory development and unification within a new foundation of physics, in which the two grand structures of field theory, the general theory of relativity and electromagnetic theory, would be united. On this kind of interpretation, one always requires an open frontier of new probing research questions which the core-context already in place is unable to cope with. Obviously, then, this could not be a motivating factor in this case.

Even though the gauge principle had been rejected in its original form, Schrödinger recognized the importance of Weyl's (generalized) gauge invariance even before the development of new quantum physics. After the development of the new quantum mechanics, Weyl lost no time to recognize it as the right core-context of the principle of gauge invariance. With some modifications (replacing the scale factor with a complex quantity and turning the scale transformation into a change of phase - a $U(1)$ gauge symmetry), he recognized that the principle provided a neat explanation for the effect of the electromagnetic field on the wave function of a charged quantum mechanical particle. The developments that followed not only changed the scene of theory development in particle physics but made it possible to understand the foundational and methodological aspects of electromagnetism and general theory of relativity even better than before.

At a higher level of methodological reflection, one might ask how far is it possible for physics to be guided by an epistemology which reflects itself so powerfully in its methodology. This question is meant to ask whether methodology itself is not open to criticism and improvement yet invariably constrained by the epistemological foundations of physics in a non-trivial and significant manner. The task of interpreting the foundational aspects of physical theory faces the biggest challenge in showing how and where exactly epistemology and methodology of science meet. Briefly, we must distinguish the dominant 20th century meta-theoretical approach from our present approach, which we may call the dynamical approach, to interpreting the foundational aspects of contemporary physical theory, where the absence or presence of their connection shows itself clearly. Notice that the meta-theoretical approach to the basic problem of understanding

---

[6]Turn to Heisenberg, W. (1984: 428; 1985:459-460; 1979: 22-41; 1969:85-101).

physical theory in its logical and methodological aspects puts an aim before it according to one's philosophical or metaphysical standpoint while assuming an unbridgeable separation between philosophy and physics or between philosophy and science (Pandit 1991: 184-197). It then asks the following questions (Popper 1983; van Fraassen 1980, 1991): How far does the given theory T succeed in fulfilling that aim? And, given T, if it has any successors or rivals, T' ... T'n, how far do they succeed or fail in fulfilling that aim? Accordingly, which is the best theory will depend on how these questions are answered by the practising physicist. Logical empiricism (Rudolf Carnap, Carl G. Hempel), scientific realism (Popper among others), constructive empiricism (B. Van Fraassen),[7] instrumentalism and structural realism are some of the best examples which follow this approach. On the other hand, the dynamical approach (Pandit 2002a, 2002b, 1982, 1991) focuses on the inner development of physical theory and theory-experiment interface within the dynamics of interaction between the foundations and the frontier of physics. As a consequence, it deals with the methodological variances resulting from this inner development. Thus, it asks how strongly are the theory's theoretical framework, its structure, problem-determination, theory-development, method, aim, strategy, instrumentation and experimental testing correlated with one another. This question opens the path to (i) finding out how within the dynamics of interaction between the foundations and the frontier of physics theory and method co-evolve and (ii) studying how methodological variances in one of them can lead to variances in the other.

5.1.2 The Methodology of Dynamic Core-Contexts of Theory Development

The philosopher Nicholas Maxwell (2006,2012a,2012b) claims that the tradition of ancient natural philosophy was lost in modern physics when Galilei's, Huygen's and Newton's development of scientific methods gained general acceptance. At this time, natural philosophy was,

---

[7]Bas van Frassen, 1980: *The Scientific Image*. Oxford: Clarendon Press; Bas van Fraassen, 1991: *Quantum Mechanics*: An Empiricist View. Oxford: Clarendon Press. For Bas van Fraassen, science aims at empirically adequate theories. A theory is empirically adequate only if a model of the appearances can be embedded in a model of the theory. One might put it alternatively by saying that a model of the appearances is isomorphic to an empirical sub-model of one of the theory's models.

according to him, splitting into science on the one hand and philosophy on the other. Nicholas Maxwell says that "struggling with the problems bequeathed to them by Descartes and Locke", philosophers Berkeley, Hume, Kant and others "produced work increasingly remote from science". In the 19th century, the continental Naturphilosophen (nature philosophers) tried to restore the unity, doing harm to science in central Europe (see Helmholtz 1862). The tradition of the theory of science (Wissenschaftstheorie) fell in the other extreme, completely separating the content of science from its meta-theoretical treatment.

However, our own approach, core-contextual methodological structuralism, is a dynamical one (Pandit 2002a, 2002b). It is closer to the path of the great philosopher-physicists of the 19th and 20th century, as diverse in their approaches as Duhem (1906), Poincare (1902, 1905, 1909), Weyl (1918a, 1981b, 1919, 1922, 1929, 1968) and Einstein (1916, 1919, 1934a, 1934b, 1960). Moreover, our approach is equally closer to the principle of Ernst Cassirer that the methodological unity must be derived from the deeds and not from the words of the scientists (Cassirer 1937). From deep inside it, physics presents a different picture altogether. At its active frontiers, physics is always on the way to the kind of new theory which it can legitimately aim at, given the core-context of theory development. Where one and the same theory presents itself through several developmental stages, the meta-theoretical tradition of twentieth century philosophy cannot account for them. If there are achievements, the great and the mature theories inherited from the past developments, these assume a dynamic role in so far as they provide the dynamic core-context for posing significant research questions and for opening of the new frontiers. Seen from this perspective of inner development, it is characteristic of physics to keep varying its frameworks, its goals, strategies, methods and aims in accordance with the dynamics built by its own mature theories opening up challenging new frontiers. The task of interpreting the search for physical theory, the striving which keeps the physicist always on the way, always asking interesting and new probing research questions, now becomes one of correlating the methodological variances, problem-determination, theory-development, theory-experiment interface, method, strategy and aim with one another, since it is to this that physics owes its constructive, at times revolutionary, movement forwards.

The theoretical development of gauge symmetries covers over half a century of theory development. It started with the formulation of Einstein's general theory of relativity and Weyl's attempts (1918, 1919) to generalize it and thereby to find a geometric basis to unify both gravity and electrodynamics, and ended, at least for the moment, with the proof of renormalizability of non-Abelian gauge theories by G. t'Hooft and M. Veltmann (Veltman 2003). Though the original ideas of Weyl were not crowned with success, the gauge principle postulated by him became an indispensable dynamic principle of theory development in particle physics in the 1960s. As a result, the Standard Model of particle physics uses similar theories to describe the electromagnetic, weak and the strong interactions. It is here that the theory of gauge invariance provides a fundamental basis for theory-unification in particle physics.

One might ask the question what is it which had guided Weyl's programme of theory-development and unification. Besides pure mathematical elegance and simplicity, what was the chief motivation and guiding idea at work in the theory of local gauge invariance evolving from the work of Hermann Weyl? Was there a basic philosophy behind Weyl's early work? In a letter dated 10th December 1918, Weyl wrote to Einstein: (quoted in Straumann, N. 1987:418)

> I am a bit sad that you cannot submit my paper to the academy; but even more, that you do not believe in it at all. ... How shall I now help myself with the redistribution of 'Space, Time, Matter'? Of course, I shall construct the infinitesimal geometry as I did it in the *Math(ematische) Zeitschrift*, in such a way that Riemannian geometry appears as a special case of a geometry at distance (ferngeometrischer Spezialfall); I cannot do otherwise without mistreating my conscience ... Incidentally, you should not believe that I was motivated by physics to introduce into geometry besides the quadratic also the linear differential form; but I wanted really to eliminate at last the 'inconsistency' which has been always a thorn in my side, and I recognized to my own amazement: this looks as if it explains electricity. You clap your hands above your head and cry: But this isn't the way one makes physics!

Weyl's complaint in the concluding statement clearly hints at one thing: There is no one and only one way of making physics. Weyl's work provides one of the best examples of the complexity inherent in theory development in physics which cannot be understood on the basis of the categories of the twentieth century philosophy of science.[8] What is most important and striking here is the structural simplicity and dynamical complexity of science and the central role it plays in the methodology of inquiry and theory-building. Methodological structuralism, in this sense, refers to the principles of theory-development as these are reflected in scientific practice itself, taken in those aspects where theory and method, theory-experiment interface, the core-context of theory development and the frontier of discovery, evolve together as it were co-variantly. At a higher level, it also entails strong correlations between the core-contextual development and epistemological evaluation of a physical theory.

In this picture, the methodological role of the dynamic core-context in guiding theory development assumes great significance, particularly in kinds of problem-situation where guidance from the experimental frontier is absent. It plays an important role not only in testing the consistency of a new theory but in developing new candidates at the active frontiers of development. This is exactly how the scenario of theory development presents itself currently: The search for a quantum theory of gravity is on, with string theory emerging as a strong candidate under the circumstance that none of its specific predictions can be tested because of the very high energy regime being involved.

Quite irrespective of the state of the art of instrumentation and experimentation, methodological structuralism has also to do with the resolving power of a theory/core-context. It is by virtue of its resolving power that a theory is able to raise new probing research questions, particularly those questions which prepare the ground for its participation in the search for a better physical theory and possibly a new core-context. If a physical theory is to serve as the core-context of theory development, this special role depends strongly on the formalism. Without the knowledge and discussion of this formalism one misses the essential aspects of the theory.[9] Even if one agrees with the

---

[8]We have considered above (Chapter TWO, sect. 2.1) how these traditions over-emphasized the logic of science, always taking well-articulated bodies of experimental evidence and theories as given in all their completeness.

[9]A good example is the highly developed formalism of "analytical mechanics" (see e.g. the textbook of Arnold 1978). Only by taking the development of this formalism into account one can understand the eventual birth of quantum mechanics. If one asumes the paradigm of classical mechanics fixed by Newton's Principia, as T. S.

younger Einstein and the older Weyl that an elegant formalism is not yet physics, one has to recognize that many developments in physics had formal reasons behind them. The gauge principle is a good example. Most current searches for physics beyond the Standard Model are governed by the formalism.

Kuhn does, one misses an important undercurrent which creates continuity.

Chapter 6

A GENERAL THEORY OF EPISTEMOLOGICALLY EMBEDDED METHODOLOGY OF SCIENCE (by G L P)

The history of physics in our century was characterized by the interplay of two extra- ordinary developments those in experimental and in theoretical physics, independent with regard to the technique and the activity of the participating scientists, dependent of each other by a continuous dialogue. The experimental development has penetrated into quite new fields and has produced a great number of new and unexpected phenomena and last but not least most important practical applications, the theoretical development has given a new meaning to such old and fundamental concepts as space, time, state, smallest unit of matter. I am afraid that it might take another century, before one has become really well acquainted with all this new scientific material and its practical, political, ethical philosophical consequences. But this will be the task for the younger generation (Heisenberg1985, Band III, 1969-1976, S. 459-460).

6.1 Rules of Method and Rules of Strategy in Science

We want to argue for the epistemologically embedded methodology of science, the thesis which is originally formulated and defended in (Pandit 1982, 1989, 1991, 1995, 2002a, 2002b). Therefore, at the very outset, we want to state clearly what kind of approach to epistemology and methodology we intend to avoid. No doubt, there have been certain major developments in the field in the twentieth century. Karl Popper developed a kind of evolutionary epistemology which places knowledge in the biological context of Darwinian theory of evolution by natural selection. Although he also calls it objectivist epistemology, the traditional epistemology being clearly subjectivist in its conceptions of knowledge and rationality of knowledge, his approach assumes relevance of biology for problems of knowledge. For example, Popper has suggested how the extension of Darwinian paradigm to epistemology by thinking of scientific theories as tentative *trial solutions* of tentatively formulated problems that are subject to error elimination by experimental refutation results in a better understanding of how scientific knowledge can grow. Thus, scientific knowledge may be viewed as a problem-solving structure with an evolutionary strategy of trial-and-error-elimination. But it is here that evolutionary epistemology shows its greatest drawback. It attends to the questions of strategy more than those of method. In nature, it may be a successful strategy for it to trigger variations among species/populations which it then selects for survival and reproduction. In the field of scientific knowledge, a similar strategy would not be adequate enough to trigger growth of knowledge and scientific change. Popper himself recognizes that objective knowledge, which he calls World3, is autonomous. And yet he fails to recognize that there is a need for a biologically neutral account of how the World3 changes while it interacts with the World2 of human minds and the World1 of physical objects (Pandit 1982, 1991). We believe that Popper's evolutionary epistemology has made it easier and simpler for W. V. O. Quine and others to propose yet another radicalization, viz. the idea of epistemology naturalized, which is now in fashion. Its basic idea is to develop theory of knowledge as a causal inquiry into the mechanism of how knowledge or a system of beliefs is produced - as a proper branch of biology and psychology (Pandit 1971b, 1982, 1991). We shall refrain here from making any detailed comment on this kind of proposal. Suffice it to say that it robs epistemology of its place as the very core of philosophy and of its normative and dynamical role in the sciences.

But there is another kind of development which views theory of knowledge as sociology of knowledge, interested in science as a social activity. For example, the work of Thomas Kuhn (1962, 1970, 1977, 1993) has much to do with the strengthening of this kind of view. Who would deny that sociological factors play important role in the functioning of modern science. If there is a trivial truth about science it is the truth that science is a social activity, or an activity which is rational in so far as the scientists employ the appropriate means to achieve the appropriate ends. However, with Steven Weinberg, we want to reject the view that the methodology, or epistemology, of science as a normative enterprise has something important to do with the sociology, or even psychology, of science. While rejecting the idea that 'our theories are not much more than social constructions, as supposed by some radical commentators on science, such as Pickering, the author of a book entitled *Constructing Quarks*, Steven Weinberg (1997, p. 42) quite admirably says that

> "We know of course that science is a social activity". As Latour and Woolgar commented after observing research in biotechnology, "The negotiations as to what counts as a proof or what constitutes a good assay are no more or less disorderly than any arguments between lawyers and politicians". But the same could be said about mountain climbing. Mountain climbers, like biochemists or lawyers, may argue over the best path to the peak, and of course these arguments will be influenced by the traditions of mountain climbing and the history and social structure of the expedition. But in the end the climbers will either get to the peak or they will not, and if they do get there they will know it. No mountaineer would write a book about mountain climbing with a title like "constructing Everest".

In (Pandit 1971a, 1982, 1989, 1991, 1995, 2002a, 2002b), one of the authors has proposed a general interaction theoretic framework for a realistic yet normative epistemology and methodology of science. In particular, he has argued for the methodology of theory-problem interactive systems in physics, distinguishing method in science from its strategy. It is this which will be presented in section **6.1.2** in outline only. But first we want to discuss in section **6.1.1** the very idea of a rational reconstruction of scientific rationality as pursued by the philosophers of science in the twentieth century but which is now in need of critical review.

We are here mainly concerned with scientific knowledge, scientific rationality and the methodology of science. Often the philosopher's statements about science, even about knowledge production generally, are so trivial that no scientist would take them seriously. In many cases, they are so uninteresting that no scientist would even care to take a look. This has even lead some scientists to ask why philosophy is so unreasonably ineffective in science, given the unreasonable effectiveness of mathematics in it. Since philosophy of science, even philosophy of knowledge, cannot itself be **knowledge-producing**, all rational reconstructions of science must face the following serious challenge: How can they be knowledge-advancing, or culture-enhancing, with significant consequences not only for public understanding of science or for the quality of life, or for our improved understanding of scientific rationality, but for all that which remains to be understood by the sciences themselves? In other words, how can philosopher-methodologists engage in non-trivially knowledge-advancing rational reconstructions which are compatible with a simultaneous search for wider horizons where we are encouraged to think in a style that is culture-enhancing? Can scientific rationality in the sense of the rationality, or rationalities, of scientific knowledge production be studied in an abstract and exclusive manner? Can it be reduced to an abstract category, if we keep in view the many-faceted complexity of science? Wherefrom should we draw our criteria for improvement in science, given the vast complexity of scientific knowledge production itself? If the philosopher's rational reconstructions are to be knowledge advancing, and not trivial descriptions, then philosophy must attend to scientific knowledge production in all its complexity. It must look at science not just in the context of justification or appraisal of its theories after they are tested by experiment but in the context of discovery understood methodologically.

Briefly stated, the major tasks or challenges for the philosopher-methodologist of the 21st century are, therefore (i) to rethink the nature of scientific rationality (Pandit 1988, 2007c, 2008, 2010a, 2010b), while keeping in view the complexity of science and the flawed conceptions of the past century; (ii) to bring about a shift in the rational reconstructions of science, while asking not how rational or successful it is but how life-enhancing are its advances and how culturally enhancing are its rational reconstructions; (iii) to ensure that science itself, or the humanity it represents, instead of the state, remains in charge of science, the larger issue being how to ensure that the moral progress mankind can make does not lag behind the scientific or technological progress; (iv) to resolve the issue of how humanity, or society, can deal with the ambivalence of science, not just with the arrogance of the scientists and non-scientists alike; so long as it remains

136

ambivalent, any legitimation is impossible and full of paradox; and (v) to re-integrate science and philosophy so that they may enjoy the same glory which they enjoyed once upon a time (Maxwell 2010, 2012; Pandit 2010a, 2010b) going beyond the meta-theoretical tradition of twentieth century.

Since in his challenging criticism of the tradition of rational reconstructions of scientific rationality, where change and improvement not just in science but in quality of life and cultural forms are our real concern, Paul Feyerabend raises many similar issues concerning the greatness of science, we will simply and briefly address these challenges (i-v) by focusing on his criticisms (henceforth Feyerabend's challenge) in section **6.1.1** below. Accordingly, the crucial question is this: Whether and how far it is possible to meet his challenge? We want to argue that it can be met without abandoning those traditions of rational reconstruction of science which are under review. To this end, the turns in the twentieth century conceptions of scientific rationality - notably in Ernst Mach, Moritz Schlick, Ludwig Wittgenstein, Rudolf Carnap, and Karl R. Popper - are briefly reviewed. Thus, both historically and methodologically it is shown how alternative traditions of rational reconstruction of scientific rationality may vary in their orientation and yet intimately share a *knowledge-advancing and a culture-enhancing* character. The real value of our proposal will lie in the implication it has for the *distinction* between the context of discovery and the context of appraisal or justification as it is known within the twentieth century philosophy of science, which is committed to the method of rational reconstruction of scientific rationality taken as the rationality of scientific change and scientific revolutions. As has been argued in (Pandit 2002a, 2002b), since its resolving power enables a theory T_1 to determine the dynamic core-context of development of the class of potential *candidates* for a new theory T_2, the very basic idea of this distinction seems to break down. Thus, a theory, a really good theory, plays both the roles in turns: Its role as explanation of the class of natural phenomena, or of the set of natural laws which describe them, and its role in determining the problemfield - what we have called the dynamic core-context of development of a new theory (Pandit 2002a, 2002b). By this it is not meant at all that an older theory, which is also the current theory in its field, should uniquely lead to a new theory. On the contrary, it should uniquely determine the problemfield which will form the dynamic core-context of development of a new theory. The divergence between the current theory and the new theory shows itself clearly in this: The current theory's resolving power plays a unique role in determining a new problemfield, which its explanatory power cannot cope with without complicating its basic assumptions sufficiently enough in an *ad hoc* manner. From this it should be clear that the search for a new theory T_2 is, as a rule, premised on a clearly determined problemfield, the dynamic core-context of the entire search-and-discovery procedure (Pandit 1989, 1991, 2002a, 2002b). We emphasize this aspect of science here precisely because it has been either neglected or forgotten or just taken for granted in all the fashionable philosophical/sociological accounts of scientific rationality, be they Carnapian, or Popperian or Kuhnian.

On the other hand, we believe that it is the prerogative of the epistemologist to analyse the questions of the structure of science and scientific change, and then to work out how the changes appearing at different levels of analysis may be correlated with one another. The question "What is science?" is the kind of question which we must keep asking as long as the special sciences make progress, resulting in the growth of knowledge. Our entire thinking about the special sciences, about their aims, methods and strategies cannot be restricted to the question of how successfully they have solved many, even most, of the problems which are known to us from the past history of science. The reason is very simple. A science is a search for knowledge and, therefore, an epistemic project always in the making. Those who have tried to suggest that our thinking about science should be conducted on *the* basis of its historical past alone are, therefore, in serious error. Our model of a science, such as physics and astronomy, should at any time instead reflect the *realism* of (i) its arrows of explanation converging in some single robust theory and (ii) its movement forwards where even the most mature and successful theory has the possibility of becoming a frontier-creating theory. As a project in the making, or as a search for the knowledge of the unknown, its method and its aim remain deeply embedded in the larger project of epistemology. Recent suggestions that epistemology is itself subject to some kind of naturalization, making it a branch of empirical science, only shows a lack of a proper sense of direction, after having traversed a long way since the time of René Descartes's and Immanuel Kant's inquiry into method. The error which we often commit is to remember the strategy a science can employ, after the rules of its method, to achieve certain ends but to completely forget the method itself, which is the only means of warranting one particular strategy instead of the other (Pandit1982, 1991). Are not the methodological rules of a science then those on which every generation premises the following kinds of question: What kind of theory **T** can it aim at in its special field? What kind of method can it follow in finding **T**, which it can aim at? And what kind of method of theory-choice can it employ, given **T** and its rivals in its field? A science can change its goals in accordance with its answers to these very questions. If the answer to our question concerning the very role of the methodological rules is in the affirmative, then each special science will search for its methodological rules of strategy afresh every time it asks and answers such questions. But this kind of search itself presupposes and requires rules/constraints at a higher level of generality. The rules of strategy should not be, therefore, confused with the rules of method, particularly where they are intended to serve the latter by extending them in certain preferred directions or domains (for details turn to section **6.1.2.** below).

What kinds of assumptions do the scientists accept as providing the foundations of the sciences? What do we understand by scientific rationality within physics? Does it have any connection with the physicist's search for those language forms - more precisely speaking, the mathematical formalisms - which are *knowledge-producing*? One might even ask here a more general question, given that the most scientific activity in physics is concentrated on theory-formation and experimental testing of its theories: Does scientific rationality within physics have anything to do with the physicist's search for those language forms which are creative and life-enhancing in a cultural sense? And, to pursue the subject still at a deeper level, one might then ask: Does the physicist's search for an appropriate language form for physics belong to the *context of discovery* or to the *context of justification*, or to both? What about those great *Gedankenexperimente* within physics?[1] Are they not yet another important aspect of his/her search for an appropriate language form for physical theory? We think that these questions are very deep and, therefore, very difficult to answer. However, we will here suggest a very brief answer as follows. In so far as a physicist, who is engaged in a search for physical explana-

---

[1]Think of the results of an unended debate over the EPR-Gedankenexperiment in Einstein, Podolsky and Rosen (1935) **47**: 777-780

tions, allows himself/herself to be guided by his/her own conception of a physical theory, we can find traces of the physicist's rational reconstructions of scientific rationality *within* physics. We can even study them in those aspects which are intended to be *knowledge-producing*. For example, think of Galileo's search for the *metaphoric* great book of nature written in the language of mathematics.[2] The language of mathematics is the framework within which science proposes to describe or explain nature, implying that the scientific language in question - take Maxwell's or Newton's equations as examples - is not to be understood as something which can be found in nature or taken as given in our minds *once for all*.

Even Sir Isaac Newton did not lag far behind when he said that "Science consists in discovering the frame and operations of Nature, reducing them, as far as may be, to general rules or laws - establishing these rules by observations and experiments, and thence deducing the causes and effects of things". As a student, having studied Robert Sanderson's *Logicae Artis Compendium* (Oxoniae, 1618) along with various works of Rene Descartes, each dealing with method differently, Newton himself distinguished between the method of invention and the method of doctrine, or the method of discovering knowledge and the method of presenting and teaching it. A more detailed answer in terms of the more sophisticated approaches of the 20th century physicists is called for. But we shall simply leave this for another occasion. We shall here instead focus on the philosopher's rational reconstructions of scientific rationality from outside physics. We shall consider them within the broader framework of the 20th century philosophy in the analytic tradition.

The main aim of the rational reconstructions of science attempted by the philosophers of science in the twentieth century has been the understanding of scientific rationality. The very question of the possibility of a choice between the *context of discovery* and the *context of appraisal/justification* for such a purpose, as if there were alternative ways of defining or explicating scientific rationality, seems to be still unresolved. For most of the analytic philosophers of the twentieth century - Bertrand Russell, Ludwig Wittgenstein, Rudolf Carnap, Moritz Schlick, Otto Neurath, Karl Popper, Carl Hempel, Wilfrid Sellars, Herbert Feigl, among others - the problem and the task of rational reconstruction of science evolved gradually from an early naturalistic

---

[2]See "Dedication" in Galileo (1632).

140

approach which was abandoned during the late 1930s, almost by all of them. Taking science in its universal context of appraisal and justification, scientific rationality came to be seen by the analytic philosophers more and more as a problem which has little or nothing to do with the context of discovery.

On the other hand, however, there exists a relatively older tradition of rational reconstruction, which has been recently developed into a highly sophisticated doctrine by Thomas Kuhn (1962, 1970, 1977). It is traceable particularly in the works of Ernst Mach (1905, 1976), to some extent in Pierre Duhem (1906) and others.[3] What we learn from this tradition is that scientific rationality has more to do with the history, sociology and psychology of scientific inquiry - what the analytic philosophers simply excluded by calling it the context of discovery. Thus, it is seen more and more as a function of the "disciplinary matrix of a science", to use a favourite phrase of Thomas Kuhn (1977: 293-319). In so far as this tradition allows its practitioners to transcend its naturalistic commitments, to present their most favoured picture of scientific change through the successive versions of their reconstruction and make choices as to the possible improvements of science, we can regard it as an alternative tradition of rational reconstruction of science in its own right. That it leaves out less and less of all that which is commonly regarded as history of scientific practice simply makes it far less abstract, yet far less challenging an enterprise than its more logically oriented anti-naturalistic alternative. In either tradition, however, there must be rules or criteria to judge whether an item belongs to the internal or to the external history of science. The most important questions which we will be concerned with in what follows are these: Is there a meeting point between inquiry into the context of discovery and inquiry into the context of appraisal/justification? Wherefrom do we pick up the criteria for improvement of scientific knowledge? Can we pick them from the scientific practice itself? Or can we pick them from the philosophy and the methodology of science, or from the convergent human interests which can be found in all cultures? Or from different rationalities that are specifically dependent on the individual cultures of different societies in the world?

At the very outset we would like to point out that no philosophical tradition of rational reconstruction of scientific rationality can

---

[3]All references are to the English translation: Mach 1976, with an Introduction by Erwin Hiebert.

be expected to be *knowledge-producing* in the precise sense in which a physical theory and its mathematical formalism are required to be knowledge-producing. This provides us important clues as to our next best expectations in this context. We just indicate them by asking the following question: How might we defend such traditions of rational reconstruction not only as *knowledge-advancing* but as *culture-enhancing*? We could not ask this question without presupposing a possible meeting point between different philosophical traditions of rational reconstruction of science. Let us assume that there is a meeting point between rational reconstructions of type (**A**) which are being articulated in the contexts of discovery and rational reconstructions of type (**B**) which are being articulated in the contexts of appraisal/justification (the latter being dependent on evaluation of various kinds of evidence relevant to theory-choice in science). The two types of rational reconstructions (**A**) and (**B**) are understood here as follows:

> (**A**) On the one hand there are rational reconstructions on the historical-genetic scale of development of scientific ideas, theories and frameworks, taking into account the disciplinary matrix of a science, the context of discovery, and what Kuhn calls lexical structures representing the primitive similarity/difference relations that are learned as part of professional science education. These structures are the source of the taxonomy/ontology of the scientific community, which undergoes change in scientific revolutions (Kuhn 1977, pp. 293-319; 1993: 311-341).

> (**B**) There are, on the other hand, rational reconstructions on the methodological scale of rational theory-choice, traditionally known as inquiry in the context of justification. The question arises: How can our rational reconstructions of science be not just *knowledge-advancing* but *culture-enhancing* if not *knowledge-producing*? How can the criteria of improvement of scientific knowledge be drawn not just from science itself as viewed through our rational reconstructions but from wider horisons of the *cultural matrixes* (**C**)?

If science is context-dependent, how do we build bridges between

(**A**)-(**B**) and the *cultural matrices* (**C**)? The variable *cultural matrices* are the variable contexts from which something new can develop. These questions raise their head because biological evolution cannot, we believe, provide norms for the growth of scientific knowledge, all of Popper's arguments not-withstanding.[4] Rational reconstructions of type (**B**) generally aim at those language forms for science which can be *knowledge-advancing* in so far as they offer:

1. Detailed logical analyses of the rules of scientific method.

2. Detailed logical analyses of the structure of scientific theory; and theoretical accounts of the growth of knowledge.

3. Historiographical reconstructions of scientific development implied by or based on (1-2).

4. Criticism of the major methodological assumptions that are revealed by (1-2).

5. And detailed normative proposals for better alternatives to replace the older methodological assumptions or frameworks, wherever possible.

We should be thankful that the humanity has inherited very rich traditions of rational reconstruction of type (**B**) as well as type (**A**). But how do we either transcend or further improve upon them? How do we use them to solve problems of improvement of scientific knowledge within the wider horizons of *cultural matrices* of human progress? What do we learn from them, if we do learn anything at all? These questions are important in their own right. Particularly, they are important for the historian of philosophy, if the history of philosophy has a future. But they assume a special significance if we have to take Paul Feyerabend's challenge to the greatness of science and scientific rationality seriously.

In order to take a look at the twentieth century conceptions of scientific rationality, here we will not go into W.V.O. Quine's proposal for indeterminacy of radical translation (or naturalized epistemology) or its strategic uses by Kuhn (1970, 1977) for his own purposes, nor into Larry Laudan's proposal of normative naturalism (turn to section 6.1.2. below).[5] Although all of them make interesting claims about

---

[4]The argument for this view is developed in (Pandit 1982/1991), going beyond Popper (1972).

[5]See Laudan 1987: 19-31. A critical discussion of Quine's (1960) and Kuhn's (1962, 9170, 1977) views can be found in Pandit 1982 (pp. 35-37) and Pandit 1991 (pp. 55-76).

the nature of knowledge and scientific reason, which are opposed to the dominant twentieth century conceptions of scientific rationality, we find them equally less challenging in their latter aspect. We prefer, therefore, to start with Paul Feyerabend's challenge we have just alluded to above. Feyerabend (1975, 1978, 1981, 1987) poses his challenge in two steps, viz., in his critique of scientific reason and in his farewell to all the abstractions of scientific rationality. By asking what is so great about science, the former provides the first necessary step to questioning the privileged status the analytic philosophers have accorded to science as a form of life. The latter step is intended to enable Feyerabend to abandon the whole philosophical tradition of analysis in which he had himself received his early training. Thus, Feyerabend's challenge takes the following form:

> Being beneficial, instead of being rational, should be the criterion both for recognizing a cultural form and for making a cultural choice, where the quality of life and of harmony and happiness are our real concern.

Are we here really caught in the dilemma, being hinted at in Feyerabend's challenge, as if we cannot pursue *culture-enhancing* philosophizing without abandoning those *knowledge-advancing*, if not *knowledge-producing*, rational reconstructions which have landed us in *meta-theoretical* abstractions of scientific rationality, scientific revolution, freedom, creativity and so forth? We want to suggest that we should take Feyerabend seriously as challenging the very idea of rational reconstruction of science, or of scientific rationality, although it can be shown by example and by abstract argument that philosophy is not here at all caught in a delimma. In other words, Feyerabend's *challenge can be met.*

Feyerabend's challenge draws our attention to the following fundamental question: Whether there does not exist an incompatibility, or a conflict beyond resolution, between the seemingly two different kinds of enterprise, particularly when they are carried out simultaneously, viz.

1. The philosopher's resorting to rational reconstructions as a means of explicating scientific rationality by proposing those language forms for science, which are knowledge-advancing in the context of appraisal and justification of its theories.

2. And the philosopher's interest in those language forms and those types of activity which are culture-enhancing and life-enhancing.

It is necessary, we think, to answer this question in order to understand the *turns* in the twentieth century conceptions of scientific rationality, particularly as these are found in Ernst Mach and Pierre Duhem at the turn of the century and then again in Moritz Schlick, Rudolf Carnap and other prominent members of the Vienna Circle; and, finally, in Ludwig Wittgenstein and Karl Popper, among others. Taking another look at their respective conceptions, and following the course of their development historically, is probably the best way to meet Feyerabend's challenge. For going into the history of rational reconstructions of scientific rationality is as important, or as relevant, in this context as moving outside this tradition in search of the *culture-enhancing* reconstructions of different types of rationality in different societies or cultures. In final analysis, then, there may not be any serious conflict between

(i) philosophy as an enterprise interested in clarifying, or explicating, scientific rationality, even if this task has to be carried out within different cultures which do not share everything in common; and

(ii) philosophy as an enterprise interested in cultural progress, given the possibility that each culture can be judged according to its own criteria for knowledge-advancing styles of thinking and culture-enhancing styles of action and living.

How do we, then, meet Feyerabend's challenge? By abandoning the familiar tradition of rational reconstruction of scientific rationality and of the structure and growth of scientific knowledge? And, then, by favouring those types of thinking and those styles of action as well as forms of interacting with the world which may be *culture-enhancing* in some sense? Why cannot we combine them both together? Why cannot we take Feyerabend's challenge as a reminder of a past heritage, which is now completely forgotten, rather than as serious advice how to engage in the philosophical and scientific pursuits of the future? We want to argue that the best way to meet the challenge is to show that to engage in rational reconstructions of science is not essentially incompatible with a search for wider horizons where we are encouraged to think in a style which is *culture-enhancing* instead of being just *knowledge-advancing* in an abstract meta-theoretical logical tradition of twentieth century philosophy of science.

There has been an enormous influence of Ernst Mach's critical-historical-genetic method on Paul Feyerabend's own philosophy of science. As early as 1957, Feyerabend [6] proposed his pragmatic theory of observation which says that it is the *observation sentence*, and not the *theory*, which is in need of interpretation. An observation sentence is distinguished from non-observation sentences not by its meaning but by the circumstances of its production. Notice how Feyerabend quite effectively delivers through this novel genetic formulation a death blow to the theory-observation dichotomy built into the logical empiricist - notably Carnap's - rational reconstruction of science (Pandit 1971a,1971b). Notice also how it favours the methodological positions of Ernst Mach and Albert Einstein (Pandit1983). Not only this, we think that it also anticipates Quine's behaviourist philosophy of language and mind, and much of his epistemology naturalized.

Did Mach (1838-1916) himself ever oppose the idea of philosophy of science as the logic of discovery? And in its place, did he favour the idea of a psychology of inquiry? The methodological thinking at the beginning of the twentieth century appears to be too complex to be easily fitted into any of the simpler alternative frameworks accessible at that time (Pandit1971b/1983). This is also borne out by Mach's (1883, 1896, 1905, 1921) creative work on epistemological and methodological problems of science. In the year 1905 Mach published *Erkenntnis und Irrtum: Skizzen zur Psychologie der Forschung.*[7] The following year saw the publication of Pierre Duhem's (1906) important work in the same field, entitled *La theorie physique: Son object et sa structure.* Finding himself in fundamental agreement with Duhem

---

[6]Detailed arguments are to be found in Feyerabend 1957: 121-130; and Feyerabend 1965: 145-218. At one place Feyerabend 1957 (p. 129) argues: "Within certain schools of philosophy, it was, and still is, fashionable to distinguish the level of everyday experience (or the 'observation-language', or the 'everyday-language') from the theoretical level, and to assume that the transition from the first level to the second level is totally different from transition between parts of either the first, or the second level. This view is a generalization of the 'orthodox' view about the relation between classical mechanics and quantum mechanics and it may therefore be called 'scientific'. But this only shows that nowadays scientists are committing a mistake which so far philosophers (notably positivistic, or 'scientific' ones) had the privilege to commit alone ... quite in general, the everyday level is part of the theoretical level rather than something completely self-contained and independent; and this suggestion can be worked out in detail and leads to a more satisfactory account of the relation between theory and experience than is the account given by Carnap, Hempel and their followers on the one side, and some contemporary British philosophers on the other." My detailed discussion on Feyerabend, Popper and Carnap in this very context can be found in Pandit 1971a, 1971b.

[7]All references here are to Mach 1976.

(1906), Ernst Mach[8] was not alone in favouring the historical-genetic method of presenting physical theories in their contexts of discovery. Of course, as disciplines, epistemology and methodology enjoyed lives of their own. But they did so as part and parcel of psychology - and probably biology - of inquiry. Thus, it would not be correct to say of Mach that his thinking about science was caught in abstractions of scientific rationality, once conceived in terms of a psychology of inquiry. Nor would it be fair to suggest that in his entire conception, methodology as the logic of discovery remained altogether unanticipated. He himself made it amply clear that he was not against the schematizing of the cognitive stages where it could be beneficial to further inquiry (Mach 1976: 223). Although he did express his opposition to methodology conceived very narrowly and exclusively. Thus Mach (1976: 223) wrote: "Schematizing the cognitive stages may perhaps benefit further inquiry when similar situations recur, but there can be no widely effective instructions for inquiry by formula." Both Thomas Kuhn (1962, 1970, 1977) and Feyerabend (1975, 1978, 1987) time and again return back to this important Machian theme. In a different language though, their arguments have become fairly well-known, using Quine's strategies (Pandit1982, 1991) and trying to show that there can be no widely effective methodology of rational theory-choice because of the incommensurability problem. No surprise, their views appear to pose most serious challenge to the tradition of rational reconstructions of the rationality of scientific revolutions. What tasks are entailed and what achievements envisaged by such reconstructions? Was not Mach far ahead of his times? Was he not himself responsible for leaving an important task for the twentieth-century philosophy of science community? We are here thinking of the normative enterprise of searching for those forms of scientific rationality which are *knowledge-advancing* in some sense. And it was this which was taken seriously as a task both by the members of the Vienna Circle and its main adversary Karl R. Popper. As we all know, after Einstein it was Popper (1934), more than anyone else, who took the challenge of psychology of inquiry versus logic of discovery very seriously (Pandit 1982, 1991).

Reworking the challenges which had originated in David Hume's psychological empiricism, Mach (1976: xxxii) proceeded with the historical-genetic method of his psychology of inquiry to achieve two

---

[8]Detailed discussion by him can be found in Mach 1976: Chapters VII-VIII, X-XVII, XIX, XXII, XXIV-XXV.

main goals: (a) removal of "an old and stale philosophy from science",[9] as he put it; *and* (b) philosophical clarification of scientific methodology and of the ways in which knowledge advances.[10] We believe that the aim of philosophy of science has not changed much ever since. On the contrary, it remains essentially the same. This is true of the members of the Vienna Circle generally as much as it is of others - notably Karl Popper, Ludwig Wittgenstein, Imre Lakatos, Paul Feyerabend, Thomas Kuhn, among others. All of them have aimed at the kind of scientific rationality which is *knowledge-advancing* in the sense of the Machian enterprise. In any case, the philosophical enterprise of clarification seeking to clarify the ways in which knowledge advances in science must itself be expected to be *knowledge-advancing*. Even their basic procedures seem to have remained the same: to rationally reconstruct science, or scientific change, under varying methodological assumptions. Thus, they have aimed at (i) critical appraisal of those philosophical/methodological assumptions which have been widely in vogue at different times in different scientific communities; and (ii) elucidation of better alternatives to these assumptions, in the hope that they might turn out to be *knowledge-advancing* rather than as purely descriptive accounts of actual scientific practice.

Let us consider now the conceptions of scientific rationality which we can find in Moritz Schlick and Rudolf Carnap, both of them being representative of the Vienna Circle. To begin with, there was the nucleus of the Circle formed around Moritz Schlick, which met regularly from 1923 onwards. The Circle came officially into being in the year 1929 with the publication of a manifesto entitled *The Scientific World View*. It was under Schlick's leadership that the activities and publications of the Circle resulted in the development of logical empiricism, the most influential school of analytic philosophy in the twentieth century. Before his assasination in the year 1936, bringing the Circle to an end, Schlick (1949) had published an important paper entitled 'Meaning and Verification" (*The Philosophical Review* vol. **45**, 1936).[11] Here Schlick distinguishes, for the first time, between the logically possible methods of verification and empirically possible methods of verification in order to improve upon his conception of sci-

---

[9]For his critical-historical-genetic method of reconstructing the development of physical theory, turn to Mach (1883, 1896, 1921).

[10]Mach 1976: xxxiii. See also Mach (1883, 1896, 1921).

[11]All references here are to Schlick 1949. Detailed discussions of Schlick's view are to be found in (Pandit 1971a).

entific rationality (Pandit 1971a). Thus, he aims at admitting all those language forms for science, in which a synthetic proposition satisfies a logically possible method of verification instead of an empirically possible one. Schlick wrote:

> Verifiability, which is the sufficient and necessary condition of meaning, is a possibility of the logical order; it is created by constructing the sentence in accordance with the rules by which its terms are defined. The only case in which verification is (logically) impossible is the case where you have *made* it impossible by not setting any rules for its verification (Schlick 1949, p. 155).

Logical possibility, the most crucial concept in this context, is defined by Schlick (1949, 154) thus: "I call a fact or a process 'logically possible' if it can be described, i.e., if the sentence which is supposed to describe it obeys the rules of grammar we have stipulated for our language". This is then further explained by Schlick (1949, 154) as follows: "... whenever we speak of logical impossibility we are referring to a discrepancy between the definitions of our terms and the way in which we use them... " Two chief arguments are offered by Schlick in support of his rational reconstruction of scientific rationality (Pandit 1971a). Schlick's reconstruction demands that the language forms for which we can specify logically possible methods of verification be given preference over those for which only empirically possible methods of verification can be specified. Correlating the legitimate language forms for science with the rationality of logically possible methods of verification for meaningful propositions, Schlick remarkably succeeds in avoiding the kind of consequences which one would have to face if the empirically possible methods of verification were assigned the role of a criterion of meaningfulness. Consider, on the one hand, the consequence of the degrees of meaningfulness, which has to do with our imperfect knowledge of the *laws* of nature, which, in their turn, play a central role in empirically possible methods of verification. And consider, on the other hand, the consequence of the naturalistic approach to problems of meaning, which Schlick now opposes. It is closely connected with the first consequence, as if meaning was something to be discovered empirically. Regarding both of them, we find Schlick (1949, 153-154) arguing as follows:

... since we cannot boast of a complete and sure knowledge of nature's laws, it is evident that we can never assert with certainty the empirical possibility of any fact, and here we may be permitted to speak of degrees of possibility... Any judgement about empirical possibility is based on experience and will often be rather uncertain; there will be no sharp boundary between possibility and impossibility.

... a proposition cannot be given 'ready made'; ... meaning does not inhere in a sentence where it might be discovered, but ... it must be bestowed upon it. And this is done by applying to the sentence the rules of the logical grammar of our language ... These rules are ... prescriptions stipulated by acts of definition ..."

Notice how decisive this is for Schlick in order to make a successful transition from a naturalistic approach to a rational reconstructionist approach to scientific rationality (Pandit 1971a). Already by the year 1934, Popper (1934) had in his *Logik der Forschung*[12] severely criticized the former approach, while Carnap abandoned it *first* in his book *Logische Syntax der Sprache* (1934)[13] and *then* in his essay "Testability and Meaning" (1936, 1937).[14] Thus, the entire focus of their attention shifted to various possible linguistic or conceptual frameworks (Pandit 1971a, 1991). Discussion of the practical questions of the possibility of choice among such frameworks by the scientific community brought about further development in the same direction. No surprise, this encouraged in Schlick himself an interest in those very propositions, or questions, which had been dismissed as meaningless by the members of the Vienna Circle earlier. In particular, we are here thinking of Schlick's (1949, 151-152, 160) discussion of the propositions about future on the one hand and the propositions about the immortality of the soul on the other. The interesting point which Schlick brings home to us is this. One might imagine a language - *a whole culture* - in which such propositions are quite well-known. One might be able to specify the logically possible methods of verification for them. In any case, questions of verification in this sense could even be left wide open.

---

[12]Popper1934. A detailed discussion can be found in Pandit 1991.

[13]A detailed discussion can be found in Pandit 1991.

[14]A detailed discussion can be found in Hempel 1950, 1965.

For further developments which followed, we must turn to Ludwig Wittgenstein on the one hand and Rudolf Carnap on the other. How is scientific rationality treated by young Wittgenstein in the *Tractatus Logico-Philosophicus* (1922)? The question is rarely raised. Although considerable influence of Mach over Wittgenstein cannot be ruled out, this work offers us the best example of a rational reconstruction of science in which the language and the world mark their co-presence through their structural isomorphism (Pandit 1971a, 1994). Thus, to imagine a language is to imagine the world and the text, one composing the other, as one single whole. Here one learns to sacrifice all the redundancies of natural languages on the one hand and of the meta-theory, on the other, taken in the sense of Russell's or Carnap's whole hierarchies of meta-languages (Pandit 1971a, 1994). And, as we might add, with Ernst Cassirer (1946, 28): "All theoretical cognition takes its departure from a world already pre-formed by language; the scientist, the historian, even the philosopher, lives with his objects only as language presents them to him". As similarly conceived in the *Tractatus*, we can say with Cassirer (1946, 37) that language is neither a copy nor a representation of a "definite world of facts, whose components are given to the human mind *ab initio* in stark and separate outlines". In the *Tractatus* (rational reconstruction), language enjoys a primordial presence in so far as it itself brings the world of facts (which are themselves actual *arrangements* of nameable objects) to presence (Pandit 1994). To borrow Cassirer's (1953) phrase again, the work of naming composes the fragmented sense experience as a world of facts and meanings. This aspect of *language* stands in a sharp contrast to its dynamic passage through time in Wittgenstein's *Philosophical Investigations* - what Wittgenstein there (*remark* 19) identifies as a *form of life*. But did he intend to suggest that language *has* a *form of life* of its own when he said that to imagine a language is to imagine a *form of life*? Did this not indicate a radical shift from the *Tractatus* (rational reconstruction) - from the paradigm of a molecular language form for science? Or is the latter essentially compatible with the proposal made in the *Philosophical Investigations*? As a tentative answer to this question, I want to suggest that in so far as the life-worlds in different cultures tend to converge on common human interests, there is no reason why there should be any incompatibility between:

(i) Philosophy as an enterprise interested in *knowledge-advancing* abstract principles of scientific rationality and the correlated language forms for science.

(ii) And philosophy as an enterprise interested in *culture-enhancing and life-enhancing* frameworks which focus on languages together with their forms of life or cultural matrices.

They need not be incompatible, even when pursued *simultaneously*. Let us consider whether we can find a meeting point between them in the work of Rudolf Carnap, or that of Karl Popper.

Of all the leading analytic philosophers of the twentieth century, Carnap's contribution to the clarification of the method of rational reconstruction, which he called *explication* of concepts, remains of unique importance in the history of philosophy of science (Pandit 1971a, 1971b). Working out one version after the other, Carnap (1950, 1956) arrived at a final version of his rational reconstruction in his essay "Empiricism, Semantics, and Ontology".[15] Among its most crucial features, we must keep in view Carnap's principle of tolerance regarding the alternative language forms for science, his revised principle of empiricism and his distinction between two kinds of question: those that are *internal* to a *framework of entities* in science and those that are *external* to it.[16] Carnap had already arrived at the first two principles in his *Logische Syntax der Sprache* (1934) and his essay "Testability and meaning" (1936, 1937) respectively. Let me restate his rational reconstruction in terms of the three-fold distinction Karl Popper (1972b)[17] makes between the *World***1** of physical objects, the *World***2** of human minds, and the *World***3** of objective knowledge, as if each of the *Worlds* **1** and **2** represented a framework of entities in its own right in Carnap's sense.

Central to Carnap's (1950, 1956) reconstruction of scientific rationality are what he calls *internal* and *external* questions. Internal questions are, then, those which the scientists can ask concerning the existence of entities *within* the *Worlds* **1** and **2**. Though the *external* questions too have the appearance of being theoretical questions, philosophers can ask them concerning the *Worlds* **1** and **2**, taking each of them as a *whole*. Thus, to imagine a language form appropriately correlated with the demands of scientific rationality is to imagine two kinds of discourse interplaying intimately, one with the other. On the

---

[15]Detailed discussions of Carnap's view are to be found in Pandit 1971a, 1971b and 1989.

[16]A detailed discussion can be found in Pandit 1971a, 1971b, 1991.

[17]A detailed discussion can be found in Pandit 1991.

one hand, we can speak of a theoretical discourse from within the natural sciences which is oriented to asking *internal* theoretical questions. For example, one may ask the following questions: What kinds of particles are electrons, protons, neutrons, positrons, neutrinos, photons, and other particles? Do they exist as free particles or only in confinement? How can we solve the quark-confinement problem? What laws of interaction, what invariances, or what symmetries, do they and others obey? On the other hand, we can speak of the practically oriented philosophical discourse asking *external* questions concerning the choice of the language form or the framework of entities as a *whole* - in the present case, the framework of elementary particle physics. The former seems to depend heavily on the strategy of asking *essentialistic* questions concerning the (*structure-property correlated*) nature and existence of the entity in question - on the strategy of methodological essentialism.[18] While the latter depends on methodological conventionalism - on the strategy of making the choice of a framework look as if it was just a matter of convention or practical decision without involving any issues of theoretical importance[19] But how can then the two types of discourse meet at some one definite point?

Although there is nothing in Carnap's philosophy of science which can prevent methodological conventionalism from penetrating deep *inside* the sciences, even turning the *internal* questions of discovery into the practical questions of choice,[20] we do believe that a real beginning is here made by Carnap to develop an approach to language and scientific rationality which is not just *knowledge-advancing* but *culture-enhancing* and *life-enhancing*. It is here that we come closer to what might be only hinted at by suggesting that the search for scientific rationality may be extended from language forms to the *forms of life*. Or, it may be extended even *to* that which Popper himself aimed at in extending methodological conventionalism to the very rules of method *within* the natural sciences.[21] As we have argued elsewhere (Pandit, 1991), the *World*3 of objective knowledge - the framework of theory-problem interactive systems in the sense of (Pandit 1982, 1991) which can undergo rational, though unpredictable, scientific changes - enables us to raise the status of the World **1**-and-World **2**-entities from mere objects of subjective knowledge (in the sense in which Descartes

---

[18] A detailed discussion can be found in Pandit 1991.

[19] A detailed discussion can be found in Pandit 1991.

[20] A detailed discussion can be found in Pandit 1991, 2003.

[21] In Pandit 1991, it is shown how Popper (1934, 1959) does exactly this.

had conceived them) to subjects of discovery, or to subjects of objective knowledge. What is most important is that this consequence has far-reaching applications in different fields, be they elementary particle physics, states of human consciousness and belief, paintings or works of art generally, living organisms or life generally, musical compositions and musical performances (Pandit 2003, 293-302; Hayes 2001). We must remember that it was Popper (1945, 1957) who argued from the unpredictability of the growth of objective knowledge to the idea of an *open society* where demands for change and reform can be met by resorting to piecemeal engineering. Similarly, wherever in society we are confronted with noticeable effects of our individual or collective moral failure, we should not be surprised if we find Popper's (1982, 1994) idea of an *open universe* - based on his theory of the *Worlds* **1**,**2** and **3** - offering new challenges to individual creativeness, individual freedom and creative intervention, be it in politics, in science, in technology, in art, in metaphysics or literature. But this raises deeper issues for further discussion which we must leave for another occasion.

6.1.1 The Methodology of Theory-Problem Interactive Systems

In the present state of development of the methodological debate, the relationship between epistemology and methodology remains as unclear and misunderstood as the relationship between methodology and the history of scientific practice itself. For example, Laudan (1987, 19-31) wrote his article, "Progress or Rationality? The Prospects for Normative Naturalism", declaring that it is not the articulation of the "criteria of evaluation actually employed in the greatest or most successful science" but the discovery of "the most effective strategies for investigating the natural world" which is the chief aim of the methodological enterprise (Laudan 1987, 27-28). What it needs is just "data concerning which strategies of inquiry tend to promote which cognitive ends (Laudan 1987, 28)." Methodology naturalized in this sense, claims Laudan, is an inquiry into the permissible means-ends connections, where data concerning them are drawn from the historical record of past science (Laudan 1987, 27-29). But it must be supplemented with an investigation into the nature and range of permissible ends of (all) inquiry, with an "axiology of inquiry", as he prefers to call it (Laudan 1987, 29). In a nutshell, Laudan (1987, 26) proposes "that the only important methodological question is this: given any proposed methodological rule (couched in appropriate declarative form), do we have - or can we find - evidence that the means proposed in the rule promotes its associated cognitive end better than its extant rivals ?" Our question is: how is this proposal going to change our usual picture of methodological rules in science? Do all such rules assert just the ends-means co-variances? If yes, how do they differ from the familiar forms of practical syllogism? Are not there more important questions concerning the methodological enterprise such as the following?

(1) What is the nature of methodological rules in the natural sciences such as physics and astronomy?

(2) How do the rules of method relate to the rules of strategy?

(3) What can the methodological enterprise aim at?

(4) In terms of which criteria can we test a methodology of science? And

(5) How can we go about warranting methodological proposals in science with a view to singling out one particular set for acceptance out of the alternative sets?

As we shall argue later, they do not exhaust all the fundamental questions which we might pose in order to disentangle the methodological complexities of *past* science and lay down norms for the *future* science. The task of answering the questions (1-5) does not become easier by giving up, with Laudan, the normative-transcendental commitments in favour of the naturalistic ones. Unlike Laudan, we believe that no theory of science can afford to pretend, within its legitimate scope of generality, to renounce such commitments. It can be argued that science such as physics has a future, just as it has a historical past. It carries its normative component within its frontier generating movements, forwards, from one particular state of development to the other, in so far as it must keep asking itself three types of fundamental questions: (1) what kind of theory $\mathbf{T}$' can it aim at? (2) What kind of method $\mathbf{M}$ can it employ in order to find out $\mathbf{T}$'? (3) And what kind of method M' should it employ in theory-choice, given the kind of T' which it can aim at? In between them, there is a whole cluster of unconventional, even unasked, questions. We think that they all have a bearing on the nature of the extant methodological rules of problem-formulation and theory-construction on the one hand and on the nature of strategies which the scientists might employ either to have the rules extended or changed, depending on how they perceive the scientific problem-situation at a given moment of time, on the other. It is, therefore, of fundamental importance to the methodological debate to distinguish clearly between (1) the methodological rules themselves which may operate at different levels; and (2) the strategies which the past scientists have always designed for employment either to make the extant methodological rules stable over longer periods of scientific research or to change, even to violate, them in the direction of revolutionary scientific change.

The muddled debate on the methodological enterprise we want here to focus on owes its origin to the conflation of method and strategy as in the case of Laudan's whole approach (Pandit 1982, 1991). We think that the real issues concerning the former are easily missed by failing to draw a clear distinction between them (Pandit 1982, 1991). The complexity of science, taken both as an achievement and as a search for the unknown theory, makes it necessary to draw the distinction. In the course of the historical development of a science, it is possible that its methodological frameworks, as also the larger aims of inquiry, remain unchanged over longer periods of time until a revolution occurs. But the strategies intended to realise those aims and to employ the correlated method, even to extend it in unknown territory, may keep changing from time to time. Where this has actually been the case in physics, comparing our strategies directly with those in vogue in earlier epochs, while judging the past successes of a science in the light of its successes today, may make little or no sense at all. We find it, therefore, hardly methodologically illuminating to be told that "to the extent that scientists of the past had aims and background beliefs different from ours, the rationality of their actions cannot be appropriately determined by asking whether they adopted strategies intended to realise our aims (Laudan 1987, 21)." On the contrary, exploring how incommensurable such strategies could be with one another cannot by itself warrant a sceptical stance regarding the prospects of a normative methodology (Pandit 1982). Nor could it warrant the view, held by Laudan (1987, 21), which is based on this very stance, that "our methodologies are precisely sets of tactical and strategic rules designed to promote our aims." As against this view, there is, we

156

suggest, more to the rules of method than the strategy which we may design to have them implemented, and in certain cases extended, in order to achieve certain well-recognized aims. In this sense, it is possible for one and the same methodology of science to allow us to design and employ any number of strategies in different situations, even those which are mutually incompatible.

As regards the question concerning the aim of the methodological enterprise, we find the kind of answer, such as Laudan's, quite unacceptable. We can even argue that the methodological enterprise cannot aim at those strategies, which a science may legitimately employ as subservient to its method. Nor can it aim at their efficacy in the investigation of the natural world. Within the framework of those distinctions which we are here suggesting, it can, on the contrary, aim much higher than that. Its highest aim remains, then, theory-finding on the one hand and rational theory-choice on the other, since it is these which play such a crucial methodological role in scientific change in one particular preferred direction instead of some other. Notice that no rules of inductive evidence framed on the bases of data supposedly drawn from past scientific practice can come to the scientist's help, when the methodologist asks the question: Which of the two theories T and T' is more promising in terms of the future developmental scenario in its field, or in terms of the highly risky novel predictions and novel experimental designs to test them?

There still remains one important question to ask:

> Who wants to make methodology more secure than physics or science itself? Is it the one who tests its strength by the rational-reconstructibility criterion or the one who argues with Laudan as follows?

It is inappropriate to expect our methodologies, he argues, to reveal as rational the theory-choices actually made by scientists in the historical scientific practice. The strategies of research incorporated in methodological rules can be tested instead by ascertaining whether we have plausible arguments and evidence that following them will enable us to make progress towards the realisation of our cognitive ends.

We think that the challenge is not to make methodology more secure than physics itself; it is rather to show that at any given time it is as secure as physics itself (Laudan 1987, 29). But the main intention behind the rational-reconstructibility of theory-choices, or even of the search-and-discovery procedures for finding new theories, was never exactly like that. On the contrary, it has to do with our search for the methodology of rational theory-choice which is sufficiently sensitive to the subtle distinction between the rules of method and rules of strategy of research on the one hand and the unpredictability of the growth of knowledge on the other. If the growth of knowledge were predictable, there would be no epistemic need at all to show how the theory-choices of the past science are rational. And if there was no point at all in distinguishing the rules of method from the rules of strategy in science, as we do distinguish, the entire methodological debate, together with our search for the methodology of theory-choice, would be seriously undermined. It is the confusion between the two kinds of rules which leads Laudan first to reduce method to strategy and then to claim that your methodology and mine cannot meet on a common point. And, as a consequence, he claims further that your successes in realising your cognitive ends cannot be evaluated by resort to our own strategies of research which we employ to realise our own cognitive ends. Incommensurability of our respective strategies entails incommensurability of our cognitive ends. And, this, in its turn, entails incommensurability of our respective inductive policies based on empirical evidence. Obviously, in this picture your strategy of research gives your enterprise as much security as your search for knowledge does.

Is there a clear sense in which we might speak of an aim of science, among possible alternative aims, which the scientists can choose in order to use it as a parameter of improvement of scientific knowledge and of epistemic appraisal within the culture of scientific rationality? In the received philosophies of science, the answer to this question has almost invariably been given in the affirmative. But Karl Popper has cautioned us on this issue as follows. Popper and his followers have argued that science aims at truth. The aim of science will always turn out to be something which the scientists can never hold as attainable. The rational acceptance of a theory T in its field involves, therefore, the scientist's belief/appraisal that it fulfils the aim of science better than its rivals do (Pandit 1988). And scientific progress is accordingly indicated by comparative degrees of truthlikeness of its theories **T**, **T'**, ... If the aim and method of science are so strongly correlated with each other, then Bas van Fraassen's constructive empiricism is an interesting variation on the same Popperian theme. It says that science aims at empirically adequate theories whose models fit the observable phenomena. The rational acceptance of a theory T in its field involves, therefore, the belief/appraisal that it fulfils this aim of science better than its rivals do. And scientific progress is accordingly indicated by comparative degrees of empirical adequacy of its theories **T**, **T'**... A detailed examination of the kind of understanding of science shown by this common belief in this kind of correlation between the aim and method of science will be undertaken elsewhere. Suffice it to ask here whether this is not really the best and most effective formula for trivializing the methodology of science by rendering it indistinguishable from anarchism, even a weak version of methodological relativism.

Is it a matter of chance, or accident, that there is life, intelligence, consciousness, creativity and successful science within the same universe of which, as we know, we can know so little? If it is not a matter of necessity, then we cannot understand either life itself or science by posing the following questions: (A) What does life aim at? And (B) what does science aim at? We are not suggesting at all that we cannot ask important questions of morality concerning life and science, which may have a similar form. According to the Darwinian evolutionary theory, life has a long evolutionary history which must be unfolded in order to understand its secrets and complexities. There is no question of life itself having a definite aim in terms of which we might understand its history, its problems and mysteries. On the contrary, if we put an aim before it - or before any one of its epochs - it is trivialized as if there is a formula to trigger the evolutionary processes of life according to a predictable pattern. But it is certain that life was unpredictable. Its future evolution must remain unpredictable if it is at all to have an interesting evolutionary history. Analogous considerations apply to science. *From this we conclude that it is a mistake to believe that there can be a single aim to guide science in its search for improved knowledge, although this belief is common to different approaches in the philosophies of science of the twentieth century. The idea of methodological variance implying the possibility that a science can change its aim in the light of its methodology has consequently remained unexplored* (Pandit 1982, 1991).

For example, even if psychoanalysis is not recognized as a science according to the extant criteria of demarcation, historically it has been one of its aims to get established as a science. For some Freud scholars, this still remains its cherished though unattained aim. At the same time, however, we cannot belittle those other types of approach, or interpretation, which look at the whole Freudian enterprise differently. The aim of the enterprise, if any, will be quite different for them. What is true of psychoanalysis seems also true of science in general. The aims of inquiry can vary even in science from one historically given discipline to the other, from one particular stage of development to the other. Not only can the aims of inquiry vary from one science to the other, one and the same science can vary its aims according to the challenges it faces in the form of successes and failures in its search for knowledge. *But trying to understand science by putting an aim before it is like trying to understand life simply by asking for its aim. Moreover, there may be important Darwinian aspects to science, analogous to the Darwinian aspects of life, as Popper has pointed out. But certainly there is more to it. And this is something which has to do with its methodology* (Pandit 1982, 1991).

The methodological theses (1-5 below) that we have deliberately chosen for discussion are rather somewhat complex, our own views interplaying with, even playing against, the views of other philosophers of science, particularly Larry Laudan (1987), to a great extent against Paul Feyerabend and Thomas Kuhn.[22] The question is what kind of connection can we make, or make out, between methodology of science and epistemology as the theory of objective scientific knowledge (turn to the Figures 6.1, 6.2, 6.3 below). What is it against which we can test a methodology of science? Through the rational reconstructions of the past science which it generates? Or by appealing to the optimal cases of successful scientific methodological variance in Newton, Einstein, Max Planck and others? Let us first state our view in the form of a set of theses (1-5) below:

1. **The thesis of *the epistemologically embedded methodology of science***

   We do not believe in an epistemologically, and therefore culturally and ecologically, uprooted methodology of science. We believe rather in an epistemologically embedded methodology of science. This should indicate that it is possible not only to raise certain fundamental questions concerning method, or change of method,

---

[22]See Passmore (1985), Blackwell (1984), Delacre (1984), Lan Ju (1986a, 1986b, 1988), McKinney (1993), Hayes (2001), Osbeck (2005).

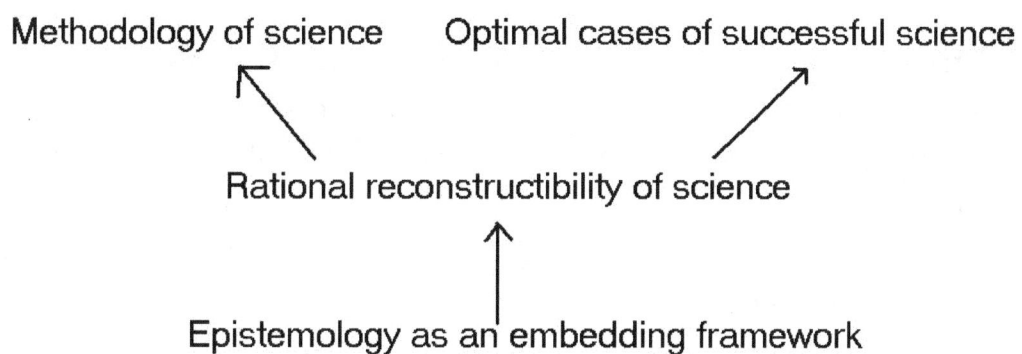

Figure 6.1: A schematic view of the epistemologically embedded methodology of science (Pandit 1982, 1991)

but also to pursue possible answers to them within the general framework of fundamental distinctions that must be drawn in the present context. Our thesis is that it is the structure and growth of knowledge which determines which methodology we choose to adopt in science. In other words, epistemology as the study of the structure and growth of knowledge, or of scientific change, is the embedding framework for the methodology of science (Fig. 6.1).

1a. **The thesis of *the methodology of inquiry as being subservient to logic of discovery*:** What is it which can warrant a methodology of science? By a methodology of science we mean here (i) a methodology of inquiry and (ii) a methodology of evaluation, the latter addressing the question how a theory is to be treated by the scientist, once it is there, and the former addressing the question of theory development. The dominant traditions in the philosophy of science of the twentieth century understand it exclusively in the latter sense. However, we think that, taken in its former sense, a methodology may be regarded as subservient to the logic of discovery that is at work in the most mature and developed sciences such as physics and astronomy. It can be shown that, in this sense, it is also closer to the actual scientific practice.

1b. **The thesis of *the distinction between method and strategy*:** What is it which can warrant a strategy of inquiry, or a strategy of evaluation? Assuming that the rules of method and rules of strategy must go hand-in-hand, it is the former which can warrant the latter in a particular problem-situation.

2. **The thesis of *the four-fold methodological distinction*:** It is necessary, we think, to draw a four-fold distinction between (i) the methodology of inquiry; (ii) the methodology of evaluation and choice, particularly when there are two or more theories to choose from; (iii) the strategies of inquiry; and (iv) the strategies of evaluation and choice.

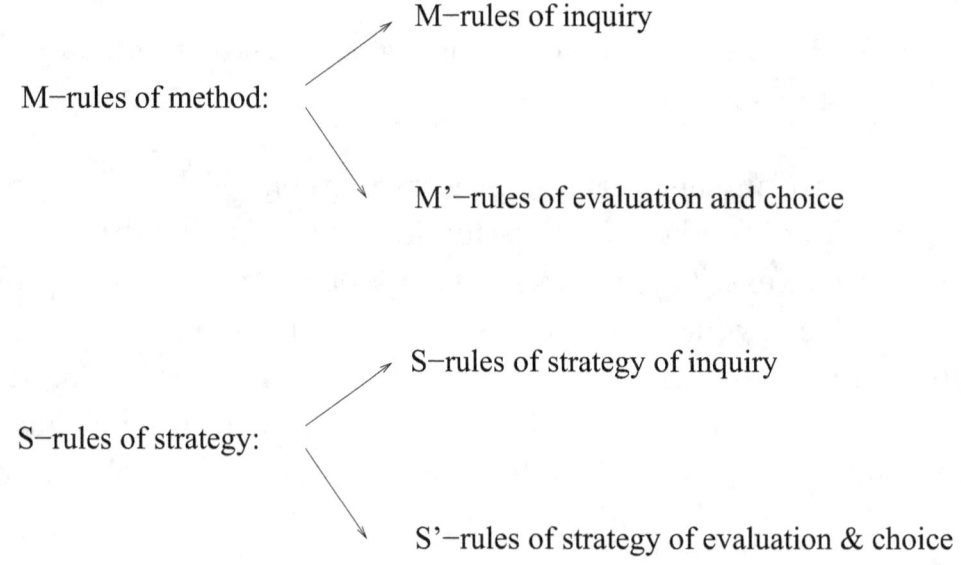

Figure 6.2: A schematic view of the methodology of science (Pandit 1982, 1991, 2002a, 2002b)

3. **The thesis of *the distinction between two kinds of rules of method (M-rules) and two kinds of rules of strategy (S-rules)*, see Fig. 6.2**

4. **The thesis of *the methodology of inquiry, if taken together with the methodology of critical evaluation, as generative of the criteria of rationality by which the individual scientist's performances/deeds, not just words, can be judged.***

5. **The thesis of *normative naturalism as trivialization of methodology*:** Against normative naturalism, we can accordingly argue as follows:

   (a) On close examination, normative naturalism (Laudan 1987) turns out to be a version of methodological relativism, even a version of anarchism.

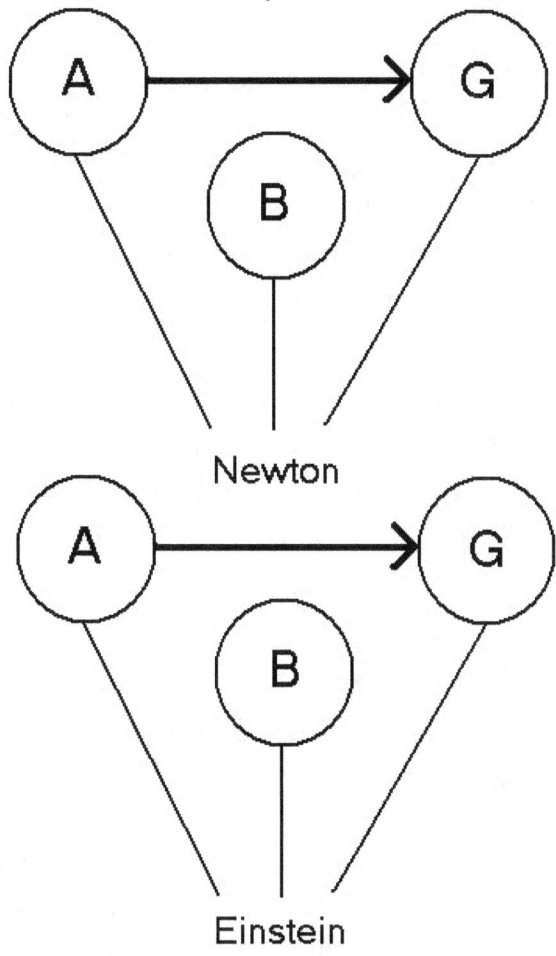

Figure 6.3: A schematic view of practical syllogistic agent rationality

(b) It conflates **M**-rules of inquiry and **M**'-rules of method with S-rules of strategy of inquiry and S'-rules of strategy of evaluation.

(c) It trivializes the methodology of science, reducing it to agent-rationality, conforming to the standard forms of practical syllogism (Fig. 6.3).

(d) It excludes what we consider to be the optimal cases of successful science (Fig. 6.6), failing to take them into account as candidates of rational reconstruction.

Thus, we want to argue against the view (see Fig. 6.3) held by Larry Laudan (1987). That is to say, we want to argue that if the methodological enterprise has an aim, it cannot be the discovery of "most effective strategies for investigating the natural world" on the basis of "data concerning which strategies tend to promote which cognitive ends." Taken together with the trivial truth that the strategies of science, those which were used in the past and those being used now, are incommensurable, Laudan's proposal can be seen as calling for an end to the methodological enterprise itself. For we find Laudan's move of remembering the strategy and forgetting the method while talking about methodology unacceptable. He makes this move as if what really matters is (i) how incommensurable any two strategies separated in historical time can be and (ii) how method is just a matter of how effective a strategy is. The basic assumption underlying this move is to view methodological rules as game-theoretic utility-structured rules of the utility-driven agent-rationality. And agent-rationality must, in all contexts, conform to the standard version of practical syllogism. Consider an agent **P** (Newton and Einstein in the Fig. 6.3, if we use a *Block Diagram* giving a schematic view of agent-rationality). His actions, in order to be rational, must conform to the standard version of practical syllogism so that it fulfills the following conditions (1-3):

(1) A historical agent P carries out an action A because (2) P desires to realise a cognitive end G and (3) P bases his action A on the belief B that doing A is the most effective strategy for realising G.

Taken together with Laudan's rejection of the methodological role of rational-reconstructibility of great science, including individual creativity, agent-rationality, in his sense, thus transferred to science is not essentially different from Kuhnian rationality of the normal scientific community. *But, as we have argued, method is not just a matter of how effective a given strategy is in relation to a particular cognitive end. Therefore, neither game theory nor standard practical-syllogistic rationality can provide us a model for the methodology of science* (Pandit 1991). Neither of them is a clue to method. Nor does the trivial truth that your (or Newton's) strategies are not my (or Einstein's) strategies undermine the normative enterprise of the methodology of science (Pandit 1982, 1991). It does not even undermine the normative methodological appraisals of great and successful science of the past by resorting to the more comprehensive standards of the present science. For that is how one can always question the received wisdom of the past and learn important lessons from history of science.

We have been examining the question whether the problem of method is just a problem of explaining the activities of the scientists with reference to standard practical syllogism as an accepted form of explanation of rational human activity. Laudan's normative naturalism answers this question in the affirmative. It is a doctrine about those strategies, sifted, on historical evidence, from the successes of the past science, which have been most effective in investigating the natural world. Laudan is not interested in the question: how are strategies themselves warranted in science, which may even be incommensurable with one another? Instead, he is interested in the question: how effective those strategies have been which Newton and Einstein, or any one else, had employed? Such an enterprise is possible, however, only if we assume, with Laudan, that there is just one and only one way of understanding the rational scientific activity of a scientist, namely by assimilating it to context-specific agent-rationality. As a consequence, we must assimilate the scientific activities of Newton and Einstein to the standard practical syllogistic form of rational activity. Rules of standard practical syllogism only tell us what the general form of those human actions is which can be assimilated to context-specific agent-rationality. But I think that it is a crucial test for any methodology of science that it does not have trivial consequences of the kind of which Laudan's normative naturalism is itself an excellent example. Who can deny that all human beings are capable of engaging in forms of rule-following activity which either conform or approximate to the standard practical syllogism. And who can deny that there are aspects of scientific activity which clearly fall in this category. But a methodology which tells us nothing more than just this must be rejected for its trivializing consequences, even the *reductio ad absurdum* consequences. On the contrary, what makes the history of physics interesting to the historiographers and methodologists is not the trivial truth that the members of physics community have all been the users of strategies dictated by their aims, interests and beliefs. Study of human behaviour, including that of the physicists, in its standard practical syllogistic, even game-theoretic, dimensions could be a temptation for any of the empirical sciences like psychology, sociology and economics. But that was not our question, to begin with. On the contrary, our question was this: What is it which the philosopher-methodologist can aim at?

In order to answer this very question, we want to argue for the following view (Pandit 1982, 1991, 2002a, 2002b). The methodological enterprise is, we believe, a search for those kinds of correlation which can exist between the scientist's risk-taking search-and-discovery-procedures for finding out the best possible theory and the search-and-discovery-procedures for finding out the best possible method of theory-choice. This kind of search is itself based on the assumption that there are cases of optimal correlations between them in the great science of the past (Pandit 1989). For example, cases of optimal correlations can be found in the forms of scientific creativity from Galileo to Newton to Einstein to Max Planck to Werner Heisenberg and others. But this brings great science *back in*, Laudan's rejection of the possible role and influence of the methodological correlations of great science not-withstanding. In essence, the problem of method is, then, the problem of understanding how *methodological variance* is possible at different levels in the sciences, like physics and astronomy, which can hold constant their formal and epistemological constraints on method which may itself vary as and when the theory varies (Pandit 1982, 1991). The following questions are here of particular interest:

(1) How to warrant a methodology as a set of variable and yet universal **M**-rules of inquiry and **M'**-rules of evaluation and choice?

(2) What can a scientist accomplish with such rules?

(3) And how can he/she test how effective they are?

The problem of *methodological variance*, particularly in physics and astronomy, assumes a more serious form as and when the *M*-rules, or **M'**-rules, break down in particular cases because our best *strategies* to have these rules extended to the newer and more complex problem-situations themselves fail. Think of the *strategies* which were designed to explain away the negative results of the Michelson-Moreley experiment at the end of the nineteenth century, or think of the problem of the stability of the Bohr atom at the beginning of the 20th century.

Scientific knowledge can grow in unpredictable ways. And yet there has to be a direction to its pursuit, which is in no way connected with agent-rationality in the sense of Laudan (1987). The most important formal constraint on *methodological variance* can be expressed in the form of the following requirement:

A sound methodology must impart direction to scientific inquiry, in particular to its risk-taking search-and-discovery-procedures which are commensurate with the unpredictability of the growth of knowledge.

At any particular time, great and successful science of the past presents itself to us as a forward-looking project with a normative component developing within its *internal life*. The methodological enterprise itself has got to be, therefore, a normative enterprise. In other words, there have got to be risk-taking universal methodological rules at the various levels of scientific inquiry before they may be violated in particular situations and before they may be replaced by better alternatives. And they can be violated despite the best *strategies* the scientists devise to fortify them wherever they tend to be least effective.

The type of *strategies* we are here hinting at are correlated with their corresponding search-and-discovery-procedures (M-rules, or M'-rules). The correlation is so strong that the former are inconceivable without the latter. One and the same methodology may warrant different types of *strategy* for purpose of attacking the theoretical, empirical and mathematical problems of varying complexity. And they may all be mutually incommensurable. Consider how the physicists of the nineteenth century resorted to pictorial and mechanical models so frequently. But what is it then which distinguishes one methodology from other methodologies? It is not the incommensurability of their *strategies* separated over historical time which can account for their differences. On the contrary, it is the fact that a methodology is capable of warranting a *strategy* by which it is then uniquely *tested*. Here, much will depend on, of course, how complex those problem-situations are which call for a strategic fortification of the methodology in fashion at a given time.

When can we say of a given methodology, or of *methodological variance* relative to it, that it is able to impart a certain direction to scientific inquiry in physics and astronomy? Three types of consideration are of importance here: (1) epistemological embeddedness of the methodology in question; (2) symmetrical theory-appraisals by means of its M-rules; and (3) rational-reconstructibility of great and successful science of the past (Pandit 1982, 1991). A proposed methodology can be regarded as a warranted methodology only if it satisfies the basic constraints (1a) - (3a) which are correlated to these considerations: (1a) How strongly or weakly is it embedded in a sound epistemological model of the structure and growth of scientific knowledge? (2a) Whether it permits symmetrical theory-appraisals directed equally at those theories which are separated over historical time, like Galileo's, Newton's and Einstein's space-time theories? (3a) Whether it permits rational-reconstructibility of great and successful science of the past? Questions (2a) and (3a) are obviously closely connected with each other. This naturally leads us to a rejection of Laudan's assumptions as these underlie his entire approach.

What can the physicist do when the **M**-rules (search-and-discovery-procedures) break down in particular situations? He/she can ask the following three questions: (i) What kind of theory with what kind of mathematical formalism (**MF**) and physical interpretation (**PI**) - **T** (= MF. PI) - should physics aim at? (ii) What types of problems confronting physics should be solved (even in the form of *Gedanken-experimente*), given the answer to the question (i)? (iii) What kind of method of theory-choice is possible in this new type of situation? For purpose of answering the question (ii), it is most important that physics is itself viewed as a dynamic theory-problem interactive system **I (T, P)** in the following sense (Fig. 6.4):

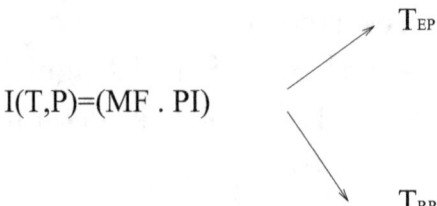

Figure 6.4: A schematic view of the methodology of Theory-Problem interactive systems (Pandit 1982, 1991). It shows that the theory **T** has both explanatory power and resolving power, the latter being responsible for the determination of problems at the frontier of discovery (Pandit 2002a, 2002b).

Einstein's methodological distinction between constructive theories and theories of principle, Bohr-Einstein debate and, more particularly, the criticism of quantum mechanics triggered by the EPR *Gedankenexperimente* are all of great relevance to answering the first two questions (i) and (ii). Both Einstein and Niels Bohr believed in scientific realism in the sense developed by the physicists themselves. But there remain unresolved between them deep differences on the questions of method, physical theory and physical reality.

The idea behind the general constraint of embedded methodology is not just to bring epistemology *back in* but to focus on its legitimate foundational role in our search for the methodology of science which can be judged both in respect of its normative and its rational-reconstructive components. In this role, epistemology is our best guide for fixing the structural identity of scientific knowledge and for searching for the best possible methodology of rational theory-appraisal. It can teach us those complexities of science which make it a project always in the making and as a successful historical accomplishment. And it can also teach us to distinguish within science its strategies from its canons for permissible strategies (Pandit 1982, 1991). The search for the methodology of rational theory-choice is a search for such *canons. Simplicity* of scientific theories as a parameter of rational theory-choice is a good example. Pursuit of scientific knowledge transcends the agent-and-context-specific pursuit of cognitive ends or human interests by means of strategies conforming to the forms of standard practical syllogism. Neither the cognitive ends nor the strategies the individual scientists may employ to realise them are important to the methodological debate. On the contrary, what is most important is the symmetry-driven aspect of science in the sense that there exist systems of knowledge in it which need to be studied as theory-problem interactive systems. The symmetrical growth of scientific knowledge from problems to theories and from theories to problems demands the recognition of these systems as problem-posing-problem-solving-systems (Pandit 1991). Science as a symmetry-driven enterprise shows a structural growth from theories to problems and from problems to theories, which follows an interactive pattern (**6.5**).

For example, the physicist at work, who is in search of the best possible theory in his field, will keep asking those three types of methodological questions (i) - (iii) which we have distinguished above. In certain essential respects, a physical theory **T** is akin to the engine at work in the railway yard. The engine in the railway yard has always two kinds of movement to distinguish it from other types of engine. First, it must move *backwards* in order to get entangled with the railway coaches which have their *destinations* already inscribed on them. But, secondly, it must move *forwards* to take *them* all to their respective destinations and, then, get disentangled from them again. From these two kinds of movement alone has the engine in the railway yard grown in time to ever new orders of structure-property correlated speed and efficiency.

$$T_{RP}(T_2) \quad > \quad T_{RP}(T_1)$$

and

$$T_{EP}(T_2) \quad > \quad T_{EP}(T_1)$$

Figure 6.5: A schematic view of the methodology of scientific appraisal of theories T_1 and T_2 at work, with T_2 having greater resolving power and explanatory power than T_1 (Pandit 1982,1991).

In a sharp contrast to our meta-methodological constraints (i) - (iii), Laudan's (1987, 21) normative naturalism can now more clearly be seen as making the following unacceptable assumptions. (a) The test of adequacy of a methodology does not consist in rational-reconstructibility of the great and successful science of the past, which includes what we have called cases of optimal correlations. (b) Necessary conditions for a rational pursuit of agent-and-context-specific cognitive ends by means of the most effective strategies are provided by forms of standard practical syllogism. (c) Rationality appraisals in science are *asymmetrical* with respect to the changing ends-means correlations. As Laudan (1987, 21) says, "our methodological rules are precisely sets of tactical and strategic rules designed to promote our aims." They cannot be, therefore, extended backwards for rational-reconstructions by asking the question: Whether the actions of the scientists of the past, with the aims and background beliefs of their own, are rational in the light of the strategies which are intended to realise our aims?

To sum up, the methodological debate about science should not be construed as a debate about the question whether in actual scientific practice every scientist is informed by a well-chosen methodology. Our rational reconstructions should tell us better about the methodological situations and goals in the sciences. The whole point of having a methodology is to be guided by a system of rules or procedures if you want to change physics by doing physics. If you do not have a clearly laid down model of how scientific knowledge can grow in physics or in astronomy, it will not be possible for you to have a normative methodology of science in these fields in the strict sense of the rules of method and rules of inquiry. The greatest methodologist, Rene Descartes, some 350 years ago, did not succeed with his methodology because he had developed a faulty theory of knowledge, its structure and its growth. Later on in the 20th century, under the considerable influence of Ernst Mach, the logical empiricists tried to develop a methodology, but they too met with failure mainly because of the flaws in their theory of knowledge. Luckily, however, outside the Vienna Circle, it was Karl R. Popper (1934, 1972b) who met with considerable success in developing a normative methodology of science based on his theory of objective scientific knowledge. But, in our view, the methodology he developed is only one-sided. It emphasizes theories in their capacity as candidates for deductive testing and rational choice for acceptance. Scientific theories, in this view, can compete for the scientist's rational acceptance, depending on how much each of them individually explains. Thus, it misses a major methodological aspect of theories, viz., their resolving power. With their resolving power, theories in science shape the active frontiers of discovery (Pandit 1982, 1991, 2002a, 2002b). And there is more to this aspect of science. Even when seen in its limited sense, Popper's methodology is traceable to Albert Einstein, to his work and methodological practice in physics. This does not necessarily mean that even Einstein's methodological insights are one-sided like those of Popper whom he influenced to a great extent.

Now, what is it which Popper's methodology misses? Like other philosophers of science, his attention remains fixed on the question how to test a theory and how to make a choice, if we have more than one theory in the field. Like most of them, Popper fails to recognize that a methodology of inquiry is as necessary for knowledge production in physics as a methodology of testing and choice is. Consequently, this implies a failure to recognize that knowledge production in the context of discovery is presupposed by knowledge production in the context of appraisal (Pandit 2002a, 2002b). Thus, a fresh approach is needed if we want to answer the question: What is science? At some point it becomes necessary to demarcate science from non-science. One may ask how easy or simple it is to do so, particularly when both the context of discovery and the context of appraisal require a fresh look from a methodological point of view. Science is not just bringing something into presence, be it a subjective experience or an objective experimental result. It is far more complex both epistemologically and methodologically. Notice, first of all, that there is an *asymmetry* between theories and problems in physics (Pandit 2002a, 2002b). As a physicist, if you have a powerful mind, you may invent as many theories as you like, both good and bad. But you can never invent problems in physics. The question arises wherefrom do then the problems come? A methodology of inquiry will be interested in finding an acceptable answer to this question. Thus, demarcating science from non-science may not be as simple a task as Popper took it to be, formulating a criterion of demarcation for its theories only. Criteria of demarcation should be formulated for both theories and problems, because of the asymmetry between them. Secondly, we know that observations cannot give rise to a theory. But a good theory can give rise to new observations. A good theory must make novel predictions. Once these have been derived from the theory, the stage will be set for making new observations that can also help in testing the theory.

The important point here is this. While it makes novel predictions, a theory in physics can function as a context for making new observations, setting up novel experiments and novel instrumentation. Most physicists would admit this. What has escaped attention and recognition is that it can also function, as it matures, as a context for inventing a new theory. In its former aspect, we are reminded of the Heisenberg-Einstein context principle that it is the theory which decides what can, and what cannot, be observed (Pandit 2002a, 2002b). Einstein's general theory of relativity and the Standard Model of elementary particle physics are good examples to illustrate the point.

6.2 An Optimal Correlations Model of Scientific Change Based on Pandit (1989)

Since it is our aim here to modify and extend our usual conception of science and scientific change, the received view as we might call it, we shall address our proposal to that effect to the scientist at work, the historian-historiographer of science, the science textbook writer and the philosopher-methodologist of science as the four types of scientific minds with their respective search-and-discovery procedures always correlated in certain ways. Against a unifying background of common interests and aims between them, where each may be engaged in his type of specialization, we wish to argue as follows:

Since any of the internal components of science can undergo scientific and rational change, with correlations showing up in other components, which may include cases of optimal correlations, the received view which implies that there is a single locus and a single direction of scientific change is mistaken and must be, therefore, rejected.

If science is always changing, with a vast network of its interactive complexities correlated with scientific creativity at various levels, and if there is more to its "internal life" than the extant models can tell, then it follows that not every state of scientific change is necessarily a state of knowledge change either in the sense of Popperian theory-change or in that of the Kuhnian conceptual change, although every state of knowledge change is a case of state of scientific change. This implies that states of scientific change are those states which need neither originate nor end invariably in states of knowledge change. And this is the first step towards a unified approach to scientific change we here aim at, viz. the idea of symmetry-driven science which has neither a preferential direction nor a preferential locus of states of scientific change but which provides for knowledge reproblematized such that knowledge change itself involves important types of correlations between theory change and problem change.

In looking for the interactive complexities underlying scientific change one should not rule out that they might be themselves subject to change in ways which may for ever remain hidden from the plain view or untouched by the narrow framework of our analysis. All cases of scientific change by a process of violation of the existing rules of scientific method, so central to Paul Feyerabend's (1978) sceptical stance, can be explained in terms of them, particularly in terms of their philosophical component. They cannot be, therefore, invoked to eliminate philosophy altogether from the scene as he has suggested by assuming additionally that science is self-sufficient and, therefore, in need of no philosophical legislation from outside (Feyerabend 1987, pp. 36, 281, 316). The same applies to the correlations of creativity emerging from such complexities in other directions. Thus it should be possible always to evaluate the extant rival philosophical accounts of scientific change by focusing, more and more sharply, on those very aspects of it which may have been left out, ignored or simply ruled out. This kind of possibility itself rests, of course, on the following important set of assumptions:

1. In the course of its historical development, unforeseen interactive complexities may develop within science taking its internal-life-components in different directions linked by significant correlations. This can happen most prominently in time of scientific revolution when new parameters in method, in problem-formulation, problem-determination, in measurement, inference and theory-construction and theory-choice assume importance to a science where they seemed to be completely absent before.[23]

2. Some of these complexities will sooner or later show themselves up in the developmental scenario of the structuration and the growth of scientific knowledge in which the scientists' favoured images of what is or is not knowledge play as important a role as those of the epistemologist himself.

3. Therefore, there will always be more to a growing science, taken in its historically, epistemologically, ontologically and methodologically interactive complexities, than what meets the specialists' eye in its extant rival interpretations and analyses. The reason why this is so is that the creativity and complexity of scientific change are better taken as correlative to each other.

To have an idea of the types of complexity which may be either at work or in progress in a particular state of scientific change, recall all those changes which have got unfolded ever since Newton changed his expository style, in which he presented the 1687 Principia, by publishing his Opticks in 1704, making it accessible to non-scientists and men of science equally. The impact has been carried in all directions, including Einstein's (1920) *Theory of Relativity, Special and General.*[24] Or consider how at any time there may be different schools of scientists within a single scientific discipline holding widely differing conceptions of their enterprise, of its aim and method, of its developmental scenario in the context of the kind of theory it aims at and the kind of method of rational theory-choice it is committed to. Thus ever since Albert Einstein and Niels Bohr debated quantum mechanics - it had

---

[23]For example, this is what the probabilistic revolution has meant to many sciences such as biology and economics. See Beaty et al. 1987, p.3

[24]For space limitations, further details relevant to this matter must be left out.

actually begun much before the EPR of 1935[25] – novel experiments have been designed one after another turning those early EPR type Gedankenexperimente into real physical ones. The latter include those recent Bell type experiments which make it possible to test the fundamental assumptions of the hidden-variable-theoretic completions of quantum mechanics against those predictions it itself makes. In this sense, then, their original debate has continued to trigger interactive complexities in physics as a whole taking it not in a single but in several directions of scientific change. The development of the problem of a consistent and an empirically adequate account of measurement-interaction in quantum mechanical context carries much significance here, since it constrains in fundamental ways any proposed alternative to the Copenhagen interpretation or its many versions.

In one of his characteristic responses to the rise of (a whole new philosophy of) quantum mechanics - to the claims and counter-claims put forth on its behalf by Niels Bohr and himself - Einstein (1954, p. 335) said: "It is open to every man to choose the direction of his striving; and also every man may draw comfort from Lessing's saying, that the search for truth is more precious than its possession." If this is a clue, the search for knowledge and the freedom (and therefore the creativity that it hints at) the individual scientist can exercise in choosing a particular direction in this search are important to scientific change. At any time, a change in direction that was chosen earlier may mean a significant change elsewhere, say in those constraints on science which may themselves be variable with scientific progress. It is quite possible that their building up is a function of the underlying interactive complexities such that at any time the scientist's search for the best possible theory in his field may be optimally correlated with the philosopher's search for the best possible methodology of theory-choice in all the fields. If this is so, the correlations between their respective search-and-discovery procedures should figure prominently in any adequate account of scientific change. Consider how in his more philosophical moments the working scientist will not hesitate to use the framework of the theory, he and his community work with, to pose new problems (with novel experimental possibilities) in the hitherto unfamiliar, unexplored or untouched problem-areas. (The great

---

[25]Einstein, Podolsky and Rosen (1935, pp. 777-780). "In all of the history of thought in recent centuries", writes Wheeler (1986, p. 361), "I know no dialogue between two greater men, over a deeper issue, reaching over a longer period of time, at a higher level of colleagueship. It extended over the twenty-eight years from 1927 to Einstein's death in 1955." See also Bohr (1935).

examples of Newton, Max Planck, Albert Einstein and Niels Bohr come readily to mind). Or, how he may suggest a novel framework for such purposes. Likewise, concerned as he is with finding his interpretative way about the historical sequence and complexity of the development of scientific ideas either within or outside the contexts of their emergence, the historian of science will not hesitate to draw our attention to the neglected aspects of a science as a living discipline or to those of a scientific figure. More significantly, he may focus on those developmental processes, either in the same field or across different fields, which relate problems to theories in a manner for which there may be no room - no readily available account - in the image of scientific knowledge dominant at a given time. In his turn, then, the philosopher of science would ask which types of change in the scientific theory-problem interaction scenario might be possible as a function of negative feedback from changes/discoveries triggered by theory on the one hand and by experiment on the other. He will not, therefore, hesitate to change and enrich his model of scientific change should there be a need to do so.

No account of scientific change which is adequate can ignore the type of mind the science textbook writer represents, nor its role in the creation of scientific traditions of rational communication, instruction and learning in a scientific discipline. Yet it may not be without reason that it should have come under heavy fire during the past two and a half decades for popularizing the following image of science: That it uses inductive methods hierarchically organizing all our knowledge as a function of cumulation, first from facts to empirical generalizations and then from one particular corpus of them to another (Kuhn 1962, pp. 1-9, 136-143). The image of science which may most easily impress him, one is likely to conclude from this, may be the most dogmatic of all possible images of it. The aim of imparting instruction in a scientific discipline by a highly ordered transmission or communication of its results and achievements may leave him with little or no option to ponder on science critically or to consider possible changes in those assumptions which the working scientist himself makes. Thus, for example, if he seriously believes that science progresses in a cumulative (= inductive) fashion, he may never question, nor recognize a need for change in, the traditional styles of instruction in a discipline which is itself undergoing change. He may fail to ask questions as crucial as the following: Whether there is need for thinking out new pedagogical sequences for teaching physics courses at introductory and advanced levels of instruction such that better altenatives to teaching them in the traditional historical sequence[26] might be explored. Whether we reject or accept this criticism, we find it more instructive to heed the great example of D. I. Mendeleev (1834-1907) who discovered the periodic system of elements while he was planning the arrangement of topics for an elementary chemistry textbook.

From the point of view of the working scientist the style of scientific change at a given time will depend upon that complex of problems, empirical or theoretical, which is in the focus of his attention at that time. It may, therefore, always vary from the making of the kind of theory he aims at to the making of the kind of experiments he is able to design, or from the articulation of the former to the refinements in the latter. In either case, it is his search for knowledge - the best possible theory - which imparts direction to scientific change. The search is constrained by the type of presuppositions and assumptions, and the search-and-discovery procedures correlated with them, which at times pull science in opposite directions. Already, we have learnt much about this aspect of scientific change from the epistemological and methodological writings of Einstein, as also from his unended search for the kind of physical theory which is at once unifying, simple, symmetrical, local and realistic in character. In his more philosophical moments, he himself had warned the philosopher-methodologist not to attend too much to the words of the theoretical physicist where one can learn much from his deeds:

> Wenn ihr von den theoretischen Physikem etwas lernen wollt über die von ihnen benutzten Methoden, so schlage ich euch vor, am Grundsatz festzuhalten: Hört nicht auf ihre Worte, sondern haltet euch an ihre Taten! Wer da nämlich erfindet, dem erscheinen die Erzeugnisse seiner Phantasie so notwendig und naturgegeben, daß er sie nicht für Gebilde des Denkens, sondern für gegebene Realitäten ansieht und angesehen wissen möchte. (Einstein 1934, p. 113).

If we take this seriously, varied aspects of the external-internal divide come to light, with the scientist's and philosopher's conceptions of the internal life of science at times sharply set against each other.

---

[26]For example, in teaching quantum mechanics in this sequence one usually puts particle physics at the end and Planck's law at the beginning.

Thus what the former admits as being internal to it the latter may recognize as being external where both may be concerned with scientific change. Consider briefly the kinds of interactive complexities one must simply rule out if one holds, with Thomas Kuhn (1962, 1970), that science (=normal science) is a disciplinary-matrix-driven puzzle-solving enterprise such that it takes a philosopher of science possibly with a Kuhnian external (= sociological) stance to tell us wherein exactly lies the locus of scientific change (Kuhn 1962, 1974, 1977). The internal-external divide takes then a paradoxical turn raising the puzzle of paradigm-confinement in the following sense. Everything "internal" to science is now fixed with reference to the paradigm such that there is no point of reference which does not fall within it either as being superseded or as itself superseding in scientific revolution. The scientist at work (= the insider) can neither operate from outside nor look for aspects not determined by the paradigm. Always confined to his (community's) paradigm, no state of scientific change, as a state of paradigm change/conceptual change which only scientific revolution can bring about, can signify any change to him. Not only does he have no access to such states, he cannot have it without simultaneous access to the alternative methodological possibilities. Therefore, he must fall back upon the philosopher's external account of scientific change, Feyerabend's self-sufficiency arguments to the contrary notwithstanding. But is this not redividing what is and what is not internal to science, according to one's convenience, by placing philosophy in remotest distance from it? Such a view, however deeply entrenched in the contemporary logical positivist practice of dividing all disciplines into first-order and second-order activities of science and meta-science, is easily refuted by the development of the physicist's own conception of scientific change in those very aspects which anticipate it in many ways (Scheibe 1988, pp. 142, 143f., 150f; Mach 1905, pp. 189, 223; Pandit 1983, pp. 393-401). No wonder, Kuhn's external sociological stance is quite silent as to which kind of stance should one adopt with respect to the history of science and science textbook writing not just as scientific disciplines in their own right but as important parameters of scientific change itself.

By the philosopher's standards, then, the dominant philosophical conceptions and assumptions which may show themselves up in the working scientist's theoretical and experimental work are to be counted as external, extra-scientific parameters of the search-and-discovery procedures that he may employ. The skeptical attack the physicists of the stature of (early) Max Planck, Ostwald, Ernst Mach and Pierre Duhem made in the last decade of 19th century against the kinetic theory is sometimes cited to illustrate how they (in this case the positivistic philosophy dominant at that time) can prejudice the scientists against their own theories at any time. We, however, think that in this particular case, as also in other cases generally, men of science of a thorough-going internalist stance like Duhem and Einstein would have sharply disagreed on this point. In their views, there is no stance outside the changing sciences which we may properly call philosophy of science. On the contrary, they are all, history of science and science textbook writing included, internally related to one another. If this were not so, it would make no sense at all to look for such relations in the contemporary science scenario, or for their optimal cases in the past, and then lament, with Marx W. Wartofsky Wartofsky (1976, p. 737), "the missing type of agreeable internal relations between philosophy and history of science", or, with Erhard Scheibe ((1988, pp. 142, 143f., 150f), a lack of the philosopher's and historiographer's appropriate response where physicists like Ludwig Boltzmann, Bohr and Heisenberg have themselves developed a fairly consistent view of scientific change anticipating those advocated by philosophers of science of later period (Mach 1905, pp. 189, 223; Pandit 1983, pp. 393-401). If the type of missing correlations here being hinted at are any clue at all, it can be shown how too much attention has been paid to the deeds of the scientist (reminding us of Einstein's advice) but too little to what he has to say about his own enterprise. The only exception to this is to be found in those interactive complexities in science where emergent correlations of creativity in different directions have shown themselves up in the work of a single man. They are the cases of optimal correlations demonstrable by the classic examples of men of science such as Isaac Newton and Albert Einstein. We think that a correct understanding of all such complexities and correlations is possible only with the help of a truly unified approach to scientific change which does not undermine the kinds of specialized investigations into science the philosopher-methodologist and the historian of science aim at. And without that it may not be so simple to correct much of the current skeptical thinking about science and about its practitioners (Feyerabend 1981, pp. 145f. and 1987, pp. 217f.)

According to the internalist stance discernible in men of science themselves, there is, then, an internal life to science, as Lakatos (1978, p. 258)called it, determined by and determining its interactive complexities arising out of these internal relations. This found its best expression in Duhem: "In a word, the physicist is compelled to recognize that it would be unreasonable to work for the progress of physical theory if this theory were not the increasingly better defined and more precise reflection of a metaphysics; the belief in an order transcending physics is the sole justification of physical theory" (Duhem 1906, p. 335). At the time when he said this, a declaration bolder than this against the externalist stance of the philosopher must have been just inconceivable. Duhem (1906, p. 334) believed that "the study of the method of physics is powerless to disclose to the physicist the reason leading him to construct a physical theory." This did not mean necessarily that it lay elsewhere quite outside physics. On the contrary, it lay in the very philosophical assumptions the physicist must make if he is to construct, as he put it (Duhem 1906, p. 324), "a system of mathematical propositions, deduced from a small number of principles, which aim to represent as simply, as completely, and as exactly as possible a set of experimental laws." Something closely similar to this is what Einstein (1954, p. 328)later on emphasized as the formal point of view of the physicist "which will sufficiently restrict the unlimited variety of possibilites" in the construction of a physical theory.

To conclude this part of our discussion, no theory of scientific change, which is adequate, can escape its own consequences for our understanding of its interactive complexities at the synchronic and diachronic levels of analysis. A theory of scientific change must be, therefore, subject to evaluation in terms of a set of conditions or requirements of adequacy along the following lines:

(i) It should be generative of non-trivial rational reconstructions, and historiographies, of states of scientific change, recovering as much structural and dynamical complexity as there may be to the growing systems of scientific knowledge. Thus it should be possible to recover from it important features of science sufficient enough to demarcate it from non-science without either trivializing its rationality or deproblematizing the structure of scientific knowledge.

(ii) It should allow as states of scientific change only those states which are accessible to the 'insider' (say, the working scientist) as much as to the 'outsider' (say, the philosopher-methodologist) such that no states of knowledge change, as special cases of states of scientific change, suffer from Kuhnian incommensurability.

(iii) It should leave enough room for correlations between states of scientific change in different directions determined by and determining the interactive complexities of science and scientific creativity at various levels.

(iv) In so far as it is these interactive complexities of its internal life which determine correlations between states of scientific change in different directions, it should itself account for the missing correlations of the type being hinted at again and again in every one-sided picture of scientific change, particularly those drawn by the philosophers and historians of science but originally anticipated by the working scientists themselves.

(v) It should, moreover, not just recover but clarify those pervasive distinctions which are presupposed by historical scientific practice at every point and with which it may be recurrently impregnated.

The requirements (i) - (iv) will be reflected, and therefore clarified, in two kinds of proposals we shall here put forth, one concerning a reproblematization of knowledge and the other an optimal correlations model of scientific change. As regards (v), I have here in mind particularly those type-distinct systems of language and science where the latter always presuppose the former. From a theory of scientific change it should be possible, therefore, to recover this together with the important fact that science is invariably embedded in but never identical with language which may go on changing without implying knowledge-change. All our uncritical assumptions of continuity, or identity, between them that we generally make are bound, sooner or later, to break down at a definite point, where it is no longer possible to look at science as nothing but another form of language and at scientific change as a kind of language change/conceptual change (as many quantum physicists have been suggesting).[27] we believe that

---

[27]This happens again and again in the physicist's and philosopher's insistence on the formalism and the logic of quantum mechanics respectively as and when their attention is drawn to its serious interpretational problems. This may be because the view that science has developed as a language among other languages, if only to enable us to speak about what Heisenberg (1958, p. 98) calls "the more remote parts of reality", is so deep-rooted in our ways of thinking that we may not find it always easy to conceptualize their growing separation from each other.

such confusion has, in fact, become a commonplace in the recent philosophy of science of Thomas Kuhn and Paul Feyerabend who draw their paradigms of scientific change from language, if only to change the subject of appraisal from scientific theory itself to the community of its practitioners holding incommensurable theories/paradigms.[28]

ls science moving more and more in the direction of a reproblematization of knowledge such that its early deproblematization, on which it has been founded by the first seekers of truth, may now be laid open to searching questions? Every attempt to define knowledge, as every attempt to assume a definition of it in search for truth, can be seen as a variation on the (universally internalized) theme of its deproblematization. Everybody has believed, until now, in the negative correlation dogma (NCD) that where there is knowledge there will be no problem, since knowledge is (assumed to be) invariably absent where there is a problem. What NCD underlines is a certain asymmetry between knowledge and problems, whether empirical or theoretical in nature. The asymmetry has always been at work in the sciences, as also in the philosophies of the sciences, where one generally starts with the constraining assumptions as deproblematizing in effect as the following:

(a) That science aims at truth (= the one and only ultimate theory Tu).

(b) That the idea of nature as being simple is best represented by the idea of knowledge as being structured by Tu.

(c) That science progresses, and knowledge grows, as it searches more and more for those theories which fulfill its aim of reducing its problems to their solutions.

(d) That through its hierarchically structured network of theories, with the earlier ones reducing to the later ones, science will have achieved its aim (= Tu) only if it has no more problems left after it has reduced them all to their solutions.

With this kind of specification, which has become an essential part of our background expectations, one looks for knowledge, either truthful or truthlike, structured by theories which have a true story of how

---

[28]lt is a characteristic feature of most rational reconstructions of scientific change that they attend just to its linguistic aspects. This is also true of Kuhn (1970, pp. 231-278), particularly where it is a matter of defending the incommensurability doctrine. See Pandit (1982, pp. lf., 14, 131-135, 164).

nature's mechanisms really work to tell. When related by appropriate reduction-chains, they must all converge on truth (= Tu). However closely we may look at the complexity of science, we are least surprised because we tend to notice and find just what we already expect to find. Thus we are unable even to perceive clearly how we have ourselves distorted what we call science or project as knowledge. Notice how even the Popperian falsifiability relation between theories and basic statements, intended to reduce the very idea of science to a convention, assumes a structuration of knowledge which is in no way different from this specification. Popper's criterion, as also his modus-tollens driven falsificationist methodology of evaluating the explanatory theories in science, is meant to be applicable only to those theories which are restricted by certain relevant logical forms. Problems, as a subject of scientific discovery, do not figure here at all. Those which do figure in the Popperian picture of science originate either in inconsistencies its theories suffer from internally or in refutations they meet in crucial experiments externally. They are essentially the same kinds of problems as meet their final solutions where later theories reduce earlier ones to themselves.

How is it possible, then, to work with an alternative specification which is not based on NCD and, therefore, deproblematizing enough to kill the developing systems of scientific knowledge before we have even touched them in all their complexity? Our answer is as follows: Let us reproblematize them by putting problems *back in* as a subject of scientific discovery. No other methodology of science we know of has yet done so. The problems we are here referring to are of a different category altogether, always there waiting to be discovered by the scientist by a process of theoretical intervention, provided there is enough theory to tell us where we might look in search of them. Once discovered, they will test the theory's realism better than its truth-likeness with which it solves the antecedently known problems which may have prompted its invention. It cannot be tested with the same force by its explanatory success, since for any number of known problems it is always possible to invent a set of theories to generate their empirically adequate solutions. Thus it could be best tested by attending to its resolving power or $T_{RP}(T)$ in the sense of Pandit (1982, pp. 12, 14, 91f., 96, 102-105, 108, 195f., 198). Where we have been accustomed to looking for theories just as solutions of problems, we can search for them as forward-looking steps to new problems. This has far-reaching consequences not just for the structuration of scientific knowledge. The dynamics built into (our) systems of problematic knowledge is seen differently as follows:

$$S(K_g)=I(T, P), (P, T),\ldots$$

Or, the growth of knowledge, as also its structuration, can be understood as a function of dynamic interaction (I) between (i) theories and problems on the one hand and between (ii) problems and theories on the other. Methodologically speaking, this entails a twin movement in a science

Backwards: P_1, P_2, P_3, $\ldots \rightarrow T_1$, T_2, $T_3 \ldots$
Forwards: T_4, T_5, $T_6, \ldots \rightarrow P_4$, P_5, $P_6 \ldots$

The methodology of epistemic appraisal to facilitate rational choice between rival theory-problem interactive systems, of the type $I(T, P)$, $I(P,T)$, will accordingly compare the explanatory power of $T_1 - T_{EP}(T_1)$ - and its resolving power - $T_{RP}(T_1)$ – with those of T_2 as follows:
$T_{EP}(T_2) > T_{EP}(T_1)$ (Backwards)
and
$T_{RP}(T_2) > T_{RP}(T_1)$ (Forwards)

Thus, scientific knowledge reproblematized will always grow symmetrically from antecedently determined problems to new theories and from known theories to new problems. Where there is knowledge there will be problems, since there will be knowledge where there are problems (Pandit 1982, pp. 102-110).

Most of the philosophical interpretations of scientific change, from the time of Galileo and Newton to our time, have been based on tacit assumptions of the structuration of knowledge on the one hand and the internal-external divide itself variously interpreted and applied to the rationality of science and its development in time on the other. Thus, according to the received view, held not only by philosophers[29] and historians (Cohen (1952, 1985) of science but many working physicists[30] themselves, scientific change essentially consists in knowledge change as a function of theory change and conceptual change such as scientific revolution may bring about. In many cases, on the other

---

[29]Popper (1934, 1959, 1963, 1972a, 1972b, 1976), Toulmin (1971), Kuhn (1962, 1970, 1974, 1977), Feyerabend (1978, 1981, 1987) and Lakatos (1971), among others.

[30]Mach (1905), Boltzmarm (1905), Duhem (1906), Einstein (1934, 1954), Planck (1949) and Heisenberg (1958,1969), among others.

hand, these assumptions reduce to the idea that any sufficiently general and abstract idea of a philosophical or metaphysical nature, whatever the manner of its influence on the development of a science, must be held as external to it and, therefore, carefully distinguished from all its internal components, relations and influences. Not only have such assumptions resulted in a misconception concerning scientific change but they stand in the way of a unified approach to it. The adequacy of the optimal correlations model of scientific change we would here like to propose only in rough outline will lie in its success in showing the states of knowledge change, which may involve correlated theory change and problem change, as special cases of the states of scientific change taking science at any time in four different but correlated directions of search-and-discovery-procedures all of which participate in its internal life (see Fig. 6.6).

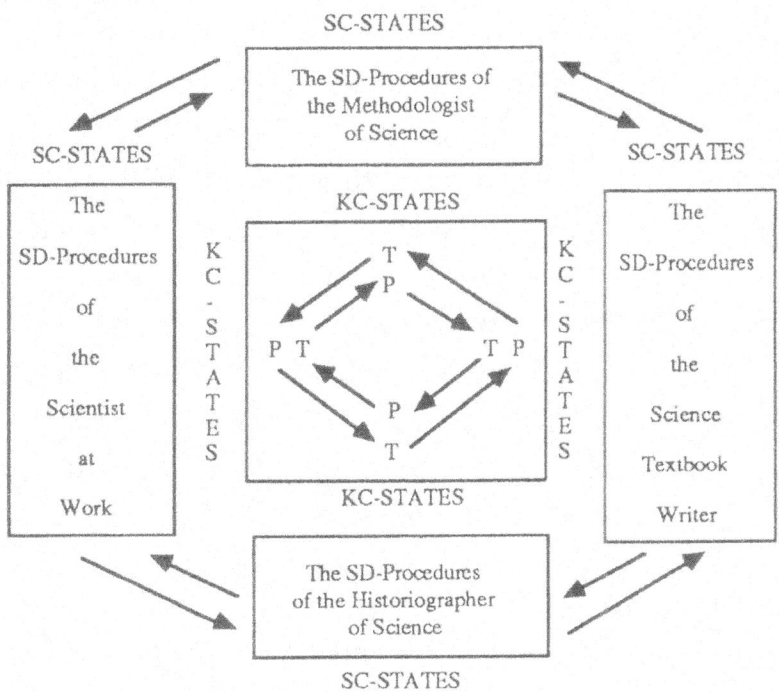

Figure 6.6: A schematic view of optimal correlations building up among the states of scientific change in different directions, using abbreviations for scientific change (SC), states of scientific change (SC-states), states of knowledge-change (KC-states) and search-and-discovery procedures (SD-procedures).

If by scientific change we may understand any change of state affecting science in any of those directions in which the working scientist, the historian-historiographer, the science textbook writer and the philosopher-methodologist respectively may at any time want to take it, states of scientific change will range over theory change/problem change as much as the changes in its historiography on the one hand and the changes in the styles of expository instruction in science as much as changes in the methodology of theory-choice on the other. The rationality of scientific change, no longer the monopoly of the philosopher of science, will lie not in any one type of state of scientific change, as against others, but in a realistic base of optimal correlations between them as being relevant to the structuration and growth of scientific knowledge. In specific cases, their character and relevance will depend on what kind of search-and-discovery procedures are either at work or in progress in each of the four possible directions. Consider, in this context, the states of scientific change in the historiography of science scenario in their correlations with those in the methodology of science scenario as a simple consequence of Lakatos methodology of scientific research programmes which tells us to look for complex research programmes, either progressing or degenerating, where we have been accustomed to look for the individual theories as candidates for true, or approximately true, propositions about the world (Lakatos 1971, pp. 91-136 and 1978, p. 212). How relevant such correlations really are to the structuration and growth of knowledge will depend on how optimal their optimality is. It takes scientific revolution to generate the kind of optimal correlations we are here referring to. Scientific revolution is itself definable[31], in this context, as that special case of change of state of science which triggers not just limited conceptual change or theory change in this or that narrow speciality but the optimally correlated states of scientific change in all the four possible directions. The strongest correlations of this kind having always flourished best in the work of a single man, their fundamental property is to build up internal constraints on science, contributing to its growing interactive complexity. Although they cannot be used to predict the future states of scientific change

6.3   The Kuhnian Paradigms vs. the Subject of Appraisal in Science

Previous versions were presented at the TU Berlin (15th December 2000), at the Institute for Physics and Astronomy (Heidelberg Graduate Days in Physics WS 8. - 12. October 2007) and at the University of Hyderabad, India ("International Seminar on Incommen-

---

[31]On Kuhnian interpretation, on the other hand, scientific revolutions are, as Feyerabend (1981, p. 147) puts it, "the outward manifestations of a change of the normal component that cannot be accounted for in any reasonable fashion."

6.3.1 Incommensurability Doctrine: Thomas Kuhn and Paul Feyerabend

At various levels, to be briefly considered below, incommensurability is claimed to be about discontinuities and non-relations between paradigms, theories, languages, perceptions, and lexical taxonomies, within the relevant fields of science. Where rival comprehensive theories in a particular domain of a science are concerned, their scientific terms are claimed to differ sufficiently in meaning, especially in reference, to render them incomparable. Despite serious problems and unresolved difficulties left over from the time of its original proposal in the 1960s, the literature on incommensurability in philosophy of science has by now assumed a life of its own, going by the recent commentaries on Thomas Kuhn (1962, 1970). A recent book devoted to "Incommensurability and Related Matters" opens with the question: What is Incommensurability? The literature in the field has grown to such an extent that some Kuhn scholars have proposed to take the discussion on incommensurability to newer heights in terms of new issues of meta-incommensurability. It has been suggested that meta-incommensurability is involved in the realism-non-realism debate (Devitt, 2001, pp. 143-157) in the sense that there are

> "several terms that change meaning when one crosses the line from realism to non-realism: namely, 'reality', 'world', 'theory comparison', 'fact', and even 'reference' itself. They purportedly refer to different things, based on the different metaphysical assumptions each party brings to the debate" (Hoyningen-Huene, Oberheim and Andersen, 1996, pp. 138-139).

We prefer to leave such complex issues to Kuhn scholars. We propose instead to take up the question "What is the subject of scientific appraisal in science?", while addressing some of the issues Kuhn and Feyerabend have raised. In particular, the two questions which we want to raise against Kuhn's doctrine of incommensurabilty are these: Whether the very idea of paradigms as effective tools or frameworks of research in science is not debatable, whichever way paradigms are conceived? Even if they are understood metaphorically, the question is whether their utility in science-studies is not highly questionable? And, if meaning-variance is part of theory-change, theory-development and theory-succession, is not the idea of incommensurable theories, theories without successors, totally inappropriate for science? We shall argue that for a physicist working at the active frontiers, paradigms not only don't exist, they make no sense as frameworks for addressing research questions of the type that are being currently asked at the frontiers of particle physics or astro-particle physics. Elsewhere (Pandit 1982, 1991), one of us has argued for this view in terms of a methodological proposal for building up a scenario of commensurable canons and incommensurable strategies in physics. Here we shall only add some further footnotes to that argument.

As early as 1962, arguing against the reductionist account of inter-theoretical relations which had been proposed by the logical empiricists, Paul Feyerabend (1981b, pp.62-69, 92-93) claimed that some successive theories T_1 and T_2 may be incommensurable, citing absence of logical relations between them due to semantic variance of scientific terms used by different theories. Around the same time, Kuhn (1962, 1970b, pp. 148-150) provided a somewhat richer account in which incommensurability ranges over paradigms such that incommensurability across paradigms percolates down to their respective components. Accordingly, incommensurability may be (i) methodological (ii) semantic or (iii) perceptual in nature, depending upon which components of the reigning paradigm and the paradigm lost are in focus. According to Kuhn (1970b, p. 103), paradigms are the source of everything that a scientific community needs: "the methods, problem-field, and standards of solution accepted by a mature scientific community". Kuhn's view goes against the traditional view that there is a uniform, invariant scientific method employed throughout science, distinguishing it from non-science. Insofar as different paradigms are themselves incommensurable, employing diverse standards of scientific appraisal and defining scientific problems differently, there will be global incommensurability including incommensurability in a methodological sense, allegedly resulting in theory-incommensurability. Again, insofar as the scientific terms employed by the scientific community change their meanings from paradigm to paradigm, there will be semantic incommensurability, allegedly rendering whole theories incommensurable. And insofar as the scientific community's own perceptions of the world depend upon within which paradigm it is working, a scientific revolution will bring about perceptual incommensurability, with scientists from the pre-revolutionary and post-revolutionary paradigms perceiving the world differently. Like Kuhn, Feyerabend (1975, pp. 23-24) too advocated the view that scientific change is characterized by changes in methodological rules and standards of scientific appraisal. In scientific change, scientific method does not stay stagnant or invariant. But Feyerabend wrongly argued from rule-violation, a science experiences in its internal life, to the denial of the applicability of universal rules of method to all science (Pandit 1982, 1991). After all, in order to be violated sometimes, a rule must be a universal rule. If a rule is not universal, there is no point in violating it.

Entangled as it is with complicated issues of meaning and reference, the subject of incommensurability in Kuhn (1962, 1970b) is a difficult subject for any significant discussion. Kuhn's views have not only seen rapid changes with changing criticisms but his shifting arguments are heavily dependent on the systematically ambiguous metaphors of a paradigm, disciplinary matrix, meaning-variance, translatability and lexical taxonomy, the lexicon being for Kuhn (1993, pp. 315, 325) "the module in which members of a speech community store the community's kind-terms" and "kind-concepts to describe and analyze the natural and social worlds". Thus, in our view, there arises a far more serious difficulty about how to understand Kuhn's doctrine of incommensurability: How far are Kuhn's (1962, 1970a, 1970b, 1971, 1974, 1977, 1981, 1983, 1989a, 1989b, 1990a, 1990b, 1991a, 1991b, 1992, 1993) changing conceptions or versions themselves meta-commensurable or meta-incommensurable?

The same question arises regarding Feyerabend (1962, 1981a, 1981b) and Kuhn (1962, 1970b). With Devitt (2001, p.144) we should ask the question: But do they "actually embrace incommensurability? Their writings on the matter are so notoriously various and vague as to leave plenty of room for disagreement over interpretation."

Originally, incommensurability was for Kuhn (1962, 1970b) a label applicable globally to what he called paradigms, with its effects allegedly percolating to various components considered above. Kuhn's doctrine took a linguistic turn with a fragmented approach to science, equating it with untranslatability: Theories T_1 and T_2 are held to be incommensurable just in case "there is no common language into which both can be fully translated" (Kuhn 1989a, 10), the two theories using different lexical taxonomies. Think of a pre-revolutionary and a post-revolutionary branch of a science. If some of the scientific terms of the former resist full translation into the lexical taxonomy of scientific terms of the latter, we have Kuhnian incommensurability of sorts. Thus, seemingly, while the focus shifts from paradigms to theories, incommensurability now applies only to a restricted set of lexical taxonomy, even going beyond the earlier concerns with meaning-variance (the noun 'mass' changing its meaning from classical mechanics to relativistic mechanics, meaning being dependent on the theoretical context). But this is not a surprising phenomenon at all. Every text book dealing with a particular branch of science normally begins by telling its readers what kinds of things, or properties, the particular science deals with. Scientific change, including a scientific revolution, is expected to bring about a change in the lexical taxonomy of scientific terms.

From this, it is quite reasonable to conclude as follows (using the abbreviation I_n (T_1 and T_2) for "incommensurability of T_1 and T_2"):

Either I_n (T_1 and T_2) is trivially true, T_2 admittedly going beyond T_1 in many significant ways in their domain , or I_n (T_1 and T_2) in its original global sense does not exist at all. In the first case, meaning-variance is one of the ways in which T_2 covers more ground than T_1. And this only enhances a healthy competition between them. In the second case, it can be argued that since there are no paradigms as tools of research in the sciences of physics and astronomy, the question of existence of I_n (T_1 and T_2) in the global sense does not arise all.

As regards the first case, it is no surprise if the meanings of scientific terms employed by scientific theories are fixed by those theories themselves. Meaning variance of scientific terms as a function of their dependence on the theoretical context does not and should not matter in theory appraisal. In other words, if meaning is contextually fixed, then it should not matter in the appraisal of T_1 and T_2 that the scientific terms employed by them differ in meaning. In fact, it does not matter at all in actual scientific practice. What matters is that T_1 and T_2 can compete in terms of their experimental consequences. For example, if T_2 performs better at the experimental frontier by receiving confirmations of its novel predictions that outnumber those made by T_1, the two theories are in healthy competition. Newton's theory of gravitation and Einstein's theory of general relativity are very good examples. Quite admittedly, there is at the semantical level a lot of meaning-variance which distances the two theories without rendering them discontinuous. This could be a source of worry only if we wanted to take them as languages and then try either to translate one of them into the other or translate both of them into a third language. We have learnt from later Wittgenstein (PU, Remarks 46-64) how absurd it is to demand reduction of one language to another language. Taking a clue from Wittgenstein, even though a scientific theory is not to be treated just as a language only, it is absurd to demand of T_1 and T_2 that they be absolutely commensurable or translatable.

On the other hand, it is rather in its latter global sense that Kuhn and Feyerabend had originally wanted to exploit its implications not only to revolutionize our understanding of science but to expose the received doctrines about science to severe criticism. Some recent studies (Pandit 1982, 1991, Fuller 2000) show that Kuhn actually produced a "profoundly conservative view of science and of how one ought to study its history". In any case, however, the doctrine of incommensurability has not evoked any significant interest among the physics and astronomy scientific communities themselves. The question is not only why this is so but also whether we should be surprised over this at all.

As far as we are concerned, leaving the difficulties just highlighted to Kuhn scholars may be the best and most wise option. We shall, therefore, limit ourselves just to the bare essentials of Kuhn's approach. In other words, we shall avoid going into the issue of how over a long period of time, while coping with his critics and admirers alike, Kuhn kept on groping for newer metaphors where his older metaphors had outlived their utility. Exchanging metaphor after metaphor, as in his "Second Thoughts on Paradigms" (1974) followed by his "Afterwords" (1993), Kuhn first moved from paradigms to disciplinary communities or matrices and then from language and translatability to lexical taxonomy (or kind-terms) and localized meaning change. One gets the impression that to the very end of his life he tried increasingly to narrow down incommensurability to a restricted class of terms, viz., lexical taxonomy of scientific terms (Kuhn 1991a, p.4, 1993, pp. 311-341), making it look as a matter of localized untranslatability.

As one of us has argued elsewhere (Pandit 1982, 1991), Kuhn's approach is distinguished by his (though not so unique) strategy of changing the subject of scientific appraisal beyond recognition. In this respect, there is no change in his later philosophy as compared to his earlier philosophy (Kuhn1962). Kuhn's incommensurability doctrine is only a consequence of his strategy. A radical shift in the subject of scientific appraisal such as this is already present in Kuhn (1962, 1970b). It is further strengthened by his resorting to the metaphors of language, translatability and lexical taxonomy in his later work in order to save the incommensurability doctrine from total disappearance. First, this strategy had brought in incommensurable paradigms. And later the same strategy brought in incommensurable taxonomic solutions in order to ensure that one is still working in a new world when a scientific revolution occurs. Along this long journey, one finds that not a single important methodological question that can be raised about scientific theory, the real subject of scientific appraisal, is either being raised or answered.

6.3.2 What does it take the Paradigms to be the Producers of Knowledge?

It was Karl Popper (1934, 1959 and 1963) who warned that we should never fall in the trap of belief-philosophies of the past and the contemporary philosophies of meaning-analysis. As regards beliefs, they are too subjective and therefore insensitive to full treatment by logical or methodological analysis. As to the meanings of words, they are far removed from the universal problems and mysteries of the universe which we would like to understand. In his book *Objective Knowledge - An Evolutionary Approach* (1972b), Popper further developed his view with considerable sophistication by proposing his important distinction between the objective World 1 (of physical things, their states and properties), the subjective World 2 (of human minds, states of belief and consciousness) and the objective World 3 (of works of art, mathematical and scientific theories). This distinction quite subtly hits at Wittgenstein's Tractatus, Logico-Philosophicus (1922) which opens with its remarks 1 and 1.1 about the world as follows:

1. The world is all that is the case.
1.1 The world is made up of facts, not of things.

Not heeding Popper's warning, Paul Feyerabend and Thomas Kuhn advocated the doctrine of incommensurability which was first based on the doctrine of radical meaning variance as an attribute of theory change in science. Part of their motivation in doing so originated in their reaction to reductionism in which the dominant philosophies of science in the twentieth century saw a powerful logical tool to understand the continuity of scientific progress from an earlier theory T_1 to a later theory T_2, where T_1 reduces to T_2 (or the language in which T_1 is formulated is translatable into the language of T_2). Such translatability is, it may be remembered, strongly built into Wittgenstein's philosophy of language in the TLP as against his PU.

Popper (1963, 1972) had also warned us as philosophers not to be misled by Wittgenstein's influential view that there are no genuine problems in philosophy to solve but only puzzles, clearly rejecting such a dogmatic view of the nature of philosophy. Before his death in the year 1994, Popper wrote his book *Alles Leben is Problemlösen* (1994), reminding us of his views which he held consistently till the very end of his life. Again, Kuhn (1962, 1974) did not heed the writing on the wall. He went so far as to identify scientific rationality with the "paradigm-driven puzzle-solving enterprise of normal science", accusing Popper of ignoring science where it is most rational. This is interesting because unlike Kuhn, Popper strongly believed in the rationality of scientific revolutions (Pandit 1982, 1991).

Unlike political economy and society, in science there is no such thing as a paradigm, in Kuhn's (Kuhn 1962, 1974) or some other sense. Of course, as we shall see later, if we are told that in science there are Kuhnian paradigms within paradigms (Kuhn 1974), the question is what are these. Again, in Kuhn's picture, there is no such thing as knowledge which is free from the web of belief, faith or conviction. Belief, faith and knowledge interplay in the production of new knowledge. Paradigms within paradigms, or whole scientific communities, are the producers of knowledge. But what is science then? On the one hand, as Kuhn's view implies, there is the paradigms-confined science (which we may simply call the problem of paradigms-confinement). On the other hand, there are Kuhnian paradigms-in-crises. Where paradigms-confined science is concerned, one cannot choose to work outside the governing paradigms. And, therefore, one cannot conceive of scientific change, since doing so would amount to thinking of working outside the governing paradigms.

Yet it may be indicative of its impact outside science that, after Kuhn wrote his book *The Structure of Scientific Revolutions* (1962), Kuhnian paradigms have come to be seen as if they dominate everywhere: In the discussions on economy and society and international relations, on the international political economy of the environment, on the market-driven economy and on globalization. What is noteworthy is that they appear to be omnipresent to those who are concerned with the clash of civilizations (Pandit 1998, 135-137). The same applies to international trade, the WTO and the UNO with all its branches spread worldwide. Paradigms can be seen in the rise, fall and resurgence of industries. One may ask whether science too has a rise, fall and resurgence as industries do. Can scientific knowledge-production be modelled after the industrial model of production of tradable goods and services? An industry running out of ideas can become sick. What about science with its Kuhnian paradigms-in-crises? To answer this question, recall how Kuhn (1962) urges us to see scientific community's skills at puzzle-solving increasingly coming under severe test where normally we would consider it a matter of scientific rationality that the theory itself is under test. The truth of the failure of the theory to explain what it is expected to explain lies actually, Kuhn's view implies, in the failure of the scientist's skills at puzzle-solving. The subject of scientific appraisal - viz., scientific theory itself - suffers a radical displacement. This raises the question whether the so-called incommensurability doctrine is not rooted in this displacement itself (Pandit 1982, 1991). Our view that Kuhn is mistaken in viewing science simply as a language among other languages is slowly and steadily gaining acceptance (e.g., turn to Nersessian 2001, pp. 275, 298, Shapere 2001, pp.181-206).

6.3.3   Kuhnian Paradigms within Paradigms: How Do our Beliefs Perform in Knowledge-Production?

Here, we are not concerned with reproduction of knowledge, which already exists in some form or other. Rather we are concerned with the problem of the production of new knowledge, given knowledge existing in the background in some form or other. Thus, the main question which we are here concerned with takes the following form: What is it which we can regard as the producer or producers of new knowledge in science such as physics? Or, how is it possible to produce new knowledge in a science with which we already have some familiarity? The most trivial answer to this question would be this: In a science such as physics, it is the scientific community as a whole which is the producer of new knowledge. But such a trivially true answer does not tell us much about the issue which is involved here. Let us then consider first what the classical epistemology can teach us in respect of this question. We think that it teaches us to separate knowledge from belief so that the producer of new knowledge in a science can be identified. This naturally leads to the view that in science it is the dominant theory, supported by experimental results, which is responsible for the production and proliferation of new knowledge. In one of its versions, this view can be expressed by saying that it is the paradigm as a whole, taken in a somewhat Kuhnian sense, which is the producer of new knowledge. But a much stronger version of this view can be found in Kuhn's own work where he distinguishes a global from a local sense of "paradigm". If we follow him and decide to view science through paradigms, within paradigms, the problem of knowledge-production as formulated here becomes highly complex. As against classical epistemology, Kuhn focuses on this complexity, reconnecting belief, faith and knowledge as if these were equal partners in the production of new knowledge. But this entails finer distinctions between different types of belief, where all of them are a causal factor in the production of knowledge (Fig. 6.7). Thus, not only belief in a theory, which is supported by experiment, but also belief in a theory which is not only new but speculative because it is not yet in a position to gain experimental support might have to be brought in as an indispensable factor in this context. In what follows, we want to focus on the following question:

> What is it which we may regard as Kuhn's achievement
> in the context of the questions raised above? Are we
> to regard his achievement as one which is free from any
> serious difficulties? If not, why?

One might wonder how many senses of "paradigm" should one try to look for, given that Kuhn (1970a, 1970b, 1974, pp.460, 463) traces a systematic ambiguity in "paradigms" in the original sense of (Kuhn, 1962). Should one talk of the paradigms-in-control-of the puzzle-solving normal science in Kuhn's sense? Or should we talk of the paradigms-in-crisis? Or should we talk of the successor-paradigm,

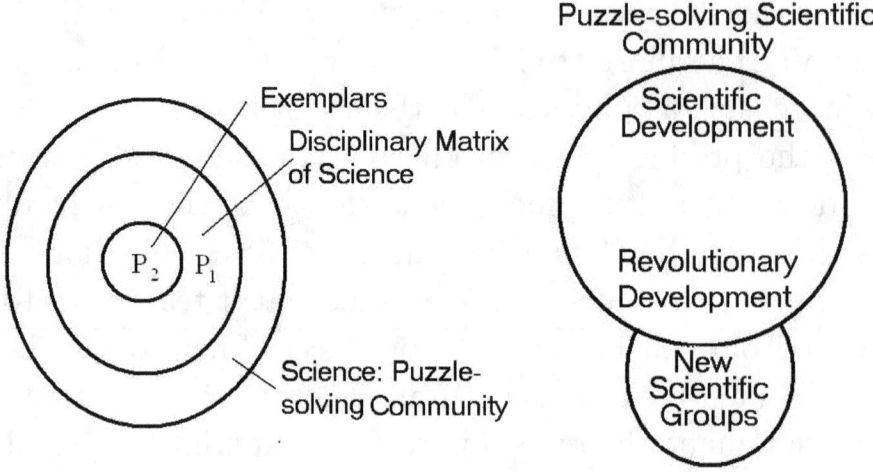

Figure 6.7: Left:A schematic view of Kuhnian paradigm-confinement of science, with paradigms (P2) within paradigms (P1). Right: A schematic view of knowledge production by speciation (formation of new scientific groupings)

just as one might talk of science, normal science and revolutionary science, one after the other? Following Kuhn, at least two senses of a paradigm are to be distinguished from each other, one of which is very broad and the other very narrow. The paradigm in the narrow sense of an exemplar is included by Kuhn as a component, or as a subset, in the paradigm understood in the broader or global sense of a disciplinary matrix (Fig. 6.7). The latter embraces all the shared commitments of a scientific community, most or all of the objects of group commitment, notably 'symbolic generalizations', 'models' (=preferred analogies) and 'exemplars' (=paradigms in the narrow sense). In Kuhn's view, understanding of all the three components of the disciplinary matrix (=understanding of normal science) is necessary to a proper understanding of how a "scientific community functions as a producer and validater of sound knowledge". In sharp contrast to this later view, Kuhn's (1962) original view had emphasized the primacy of paradigms as monolithic sources - as global determinants - of everything which a puzzle-solving scientific community needs or produces.

Most important concept in Kuhn's picture of science is the concept of normal science. More appropriately one may speak of the states of normal science which are stabilizing enough to give an identity to science. Science in this fundamental sense is interrupted by scientific changes called scientific revolutions (Kuhn, 1970b, p.92). Scientific revolutions are defined "as those non-cumulative developmental episodes in which an older paradigm (?) is replaced in whole or in part by an incompatible new one." In other words, these revolutions change the world in which a scientist works. The members of different scientific communities - those of the pre-revolutionary phase and those of the post-revolutionary extra-ordinary phase - are presented with different data by the same stimuli. Kuhn (1970b) has accused Popper of having ignored normal science: "If a demarcation criterion exists (we must not seek a sharp or decisive one), it may lie just in that part of science which Sir Karl ignores." The questions which arise here immediately are these:

1. Is the structural identity of post-revolutionary paradigm with the pre-revolutionary paradigm consistent with the claim that they are incommensurable?

2. Is it the paradigm in the global sense or in the narrow sense which is affected in the scientific revolution?

3. If a scientific revolution changes the world in which a scientist works, is it because of the paradigm-change in the narrow or broader sense, or just because of theory-change?

4. Whether the incommensurability between the paradigms is to be understood as a key to understanding scientific change? Or whether it is to be regarded as a consequence of paradigm-change?

5. Is the quantum mechanics of last 80 years normal science or revolutionary science? Is it a case of paradigm-in-charge or of paradigm-in-crisis?

Since Kuhn (1970b, p.92) allows replacement of an older paradigm by a new one in part, or in whole, when a scientific revolution takes place, questions (2), (3) and (4) are not so easy or simple to answer.

Kuhn's greatest achievement may appear to lie in displacing the scientific theory, in the standard sense in which practicing physicists understand it, as a subject of rational scientific appraisal, given his view that it is the puzzle-solving scientific community bound by a disciplinary matrix which is under test when a science is in crisis. But at the same time, he seems to favour an interpretation of science, taken in its normal and revolutionary phases, as an activity which is dominated by theory, if it is correct to identify his narrower notion of a paradigm (=exemplar) with the notion of a theory. The paradox here is this: If science is dominated by theory, it is the theory which should bear the main ordeal of the experimental test of all its predictions. But if we are told that we should think of the whole community of scientists, their skills at puzzle-solving, bound by a disciplinary matrix, as the main bearer of the ordeals of testing, then it is clear that the theory must be displaced from that important stage which it has occupied, and is still occupying, in the actual scientific practice.

In the Kuhnian normal science, conceived as a puzzle-solving enterprise strongly restricted by the disciplinary matrix, it turns out that the most important concept is that of the community/group which holds a theory and which works within a paradigm. It is not the concept of the theory itself. As Kuhn (1993, 338) puts it: "whether or not individual practitioners are aware of it, they are trained to and rewarded for solving intricate puzzles - be they instrumental, theoretical, logical, or mathematical - at the interface between their phenomenal world and their community's beliefs about it". Kuhn's account is explicitly and predominantly a sociological one. To an unprejudiced reader of Kuhn (1962,1970b), it would at first seem as if it is in his discontinuity-view of conceptual change, itself a function of scientific revolutions which result from paradigm-change, that the search for scientific rationality finally makes a contact with normal science. For, if we have to look for scientific rationality anywhere in science, we should look to scientific change taking place within it. But the Kuhnian picture is far from a simple picture. Not only does the sociological paradigm of normal science, which Kuhn brings in while putting his picture of science in place before us, block our way to distinguishing clearly that which is internal to science from that which is external to it, it threatens even our analysis of scientific change at all the relevant levels with relativistic consequences. Let us consider now what these consequences are like.

(a) Think of physicists who hold a theory T, with a critical attitude to its scope of applications and to its power for unifications among the extant theories, given that T is the kind of theory they had been aiming at. Their critical attitude to T would entail the kind of rational understanding and use of T which does not close the possibility of a relentless search for alternatives to T at any time. T may be the dominant theory in its field. Yet a critical attitude to T will be warranted by the practicing physicist's own conception of how physics can make progress by a relentless search for alternative methodological possibilities. We think that this consideration is crucial to taking a fresh look at Kuhn's conception of normal science. While working within a paradigm, do the members of the normal scientific community hold T with the same kind of understanding? If the answer is yes, then it is working with T while leaving open the possibilities of finding alternatives to T. But that is exactly what the Kuhnian picture does not allow. There is no way for the normal scientific community to conceive of scientific change from within the paradigm it is working in. In other words, neither philosophy nor methodology figures as a critical parameter anywhere in the internal life of normal science. No surprise, then, that Kuhn should choose not just a historical but a sociological turn to tell us about scientific rationality and scientific change. For a sociological analysis to be in place, it is being assumed that there is available within science itself a sociological basis to support such analysis. But we think that this does not necessarily undermine the normative enterprise of philosophy of science. On the contrary, its possibility as a meta-scientific enterprise is facilitated by the assumption that normal science itself as a first-order activity is always philosophically neutral, since philosophy does not (or perhaps does not have to) figure in its internal life. But this raises the question whether this does not entail the absence of that tension between knowledge and belief, or between knowledge and faith, on which the classical epistemology of knowledge-production has been thriving for so many past centuries.

The classical epistemology assumes a separation between them, which no definition of knowledge can be expected to ignore. But now the Kuhnian paradigm-confined science re-unites them. Interplay of belief and knowledge is not a new phenomenon unknown to physics or philosophy. Every theory that matters to physics is a subject of belief or disbelief in the community of physicists, depending on what kind of evidence they are able to gather from the experimental tests of the theory. But the Kuhnian paradigm-confined science brings into play beliefs of an entirely another kind. They are visible only in those leaps of faith which are supposed to characterize the paradigm-shifts in Kuhnian sense. "The man who embraces a new paradigm at an early stage," writes Kuhn (1970b, p. 220)," must often do so in defiance of the evidence provided by problem-solving. He must, that is, have faith that the new paradigm will succeed with the many large problems that confront it, knowing only that the older paradigm failed with a few". If the paradigm-shift - a scientist's embrace of a new paradigm in place of an older one faced with crisis - is a matter of a leap of faith, "a jump across incommensurate canyons of irreconcilable beliefs", as Caroline A. Jones (Spring 2000) has put it, then it is necessary to ask the following question: How far can we regard a paradigm as a factor in scientific knowledge-production? Understood in a global sense, Kuhn's claim seems to go even further: The paradigm as a whole must be regarded as the producer of scientific knowledge.

(b) The difficulty just considered raises a further question: Whether in reality there is no more to science than a closed society of scientists engaged in the activity of puzzle-solving in an uncritical fashion. To be sure, no historiography of science based on such a picture would stand the scrutiny in terms of the actual cases - at least the optimal cases - where the practicing scientists have all along held a definite view of how their science can make progress. It is, therefore, possible to show how Kuhn's generalizations about science are refuted by the actual historical development of science.

(c) Even if we accept the Kuhnian picture of normal science, there still remains a very serious problem of paradigm-confinement which can be formulated as follows. How is scientific change conceived as theory change and problem change, or as radical conceptual change, motivating or itself motivated by paradigm-change, possible? Since there is no occasion, nor any possibility, for the members of a normal scientific community to conceive of scientific change so long as they are working within the paradigm to which they must remain confined, how is scientific change under the Kuhnian condition of paradigm-confinement possible at all? One fine morning they may wake up and start working within a new paradigm, but in a manner that makes the earlier and the later paradigm incommensurable. That is the kind of scientific change, Kuhn tells us, which a scientific revolution brings about. But it is obviously something which no practicing scientist can either perceive or detect. It is beyond his/her powers of perception. A practicing scientist is always confined to the paradigm which is strongly restricted by the disciplinary matrix. It is rather left to the philosopher-sociologist's externalist stance to tell us about science and scientific change, the practicing scientist being all the time confined to his one and only one paradigm. Without a thoroughgoing externalist stance, therefore, it should be impossible to see clearly where scientific change has taken place, interrupting the stabilizing phases of normal science. Again, no historiography of scientific change based on this picture can stand a close and critical scrutiny. We may agree that there are external contexts of scientific discovery, closely akin to Kuhn's paradigms or disciplinary communities, that define what he calls normal science.

6.4 Effective Core-Context of Theory Development in Physics: In Retrospect

We are here concerned with the problem of how, after passing tests and becoming mature, a physical theory contributes to context-building in theory-finding and theory-testing, creating new frontiers of theory development in physics and astronomy. [32]

The leading twentieth century philosophers of physics, notably Hans Reichenbach and Rudolf Carnap, among others from the Vienna Circle, joined later on by Karl R. Popper, were pre-occupied with the task of imparting a new direction to research in philosophy. Following a programme of analysis of the logic of science, they thought that in order to carry out their programme, it was imperative to distinguish from each other two kinds of contexts relevant to science studies, viz., the context of justification and the context of discovery. Philosophy, conceived as the logic of science, constituted a field of science study focusing on the scientific theory exclusively in the context of justification. Taken in this context, what was philosophically relevant to serious science study, they argued, were the questions of rationality of theory-testing and theory-choice in science. Thus, the philosophical study of science in the context of justification assumed the unique significance of addressing the task of meta-theoretical formalization of the rules of scientific rationality. Within science as a settled discipline, with the help of such rules, theories as finished products of the process of discovery could be tested and then evaluated for rational choice, taking into account elaborate bodies of empirical evidence. Thus, only theories themselves, taken in their proper context of justification or confirmation in science, could be taken as the legitimate subjects of philosophical analysis by rational reconstruction.

In this way, it was made quite clear by the philosophers themselves that philosophy had nothing to do with science studies in the context of discovery, i.e., with science-in-context. On the other hand, all that

---

[32]For detailed arguments see G. L. Pandit (2002a, 2002b).

which fell within science-in-context was to be excluded from philosophy. For example, think of a context in which a scientist is able to conceive of new ideas which we have never heard of before. The creative act of conceiving novel ideas in science belonged, according to the analytic philosophers of science, to science study in the context of discovery. Being external to science as a settled system of knowledge, and there being no logic of discovery, the context of discovery could not, they argued, form a legitimate subject of philosophical analysis by rational reconstruction. This particular concern with exclusion of the context of discovery from philosophy of science, in particular from the philosophy of physics of the twentieth century, shaped quite decisively their entire programme of philosophical analysis of scientific knowledge as well as scientific changes taking place in physics and other natural sciences.

After a century of philosophical opposition to science-in-context, i.e., to science study in the context of discovery, the question today arises whether in doing so they did not throw away the baby along with the bath water. It is noteworthy that in most dominant traditions of the twentieth century philosophy of science, science-in-context refers to something which lies quite outside the logic of science. While ruling it out as a possible subject of philosophical rational reconstruction, the philosophers did not even consider context-building as an important dynamic factor of scientific development, playing an important role within the natural science itself. Among the great physicists of the twentieth century, notably Einstein and Heisenberg, probably no one thought of what Reichenbach, Carnap and Popper, or the logical positivists generally, were up to when they ruled out any concern with science study in the context of discovery. It is against this background that we would like to focus on Heisenberg-Einstein dialogue on physical theory going back to early 1926. When the quantum core-context building appeared on the horizon of physics, they began the dialogue with a view to resolving the problem of reconnecting observation with theory. It is, therefore, quite pertinent to pose the following question: Whether this is not the right occasion to take another look at the Heisenberg-Einstein context principle as it emerges from their dialogue on physical theory? Whether we should not explore the broader issues concerning physical theory which are still waiting for serious methodological attention, context-building in the dynamics of theory development in physics being one of them?

We are here referring to their context principle which they formulated as follows: "that it is the theory which decides what we can observe." With this principle, Heisenberg and Einstein reconnected observation with theory, and re-integrated the frontier of discovery in physics with the core-context of development, thereby restoring to physics and astronomy what they had been robbed of under their earlier highly restrictive doctrines. As we shall notice in what follows, their earlier doctrines turn out to be a variation on the Ernst Mach's foundationalism. They had required that a theory must be founded on observable magnitudes alone, that in its entire structure it must include nothing but observable quantities.

To pursue our discussion further, it is very important to ask the following two questions:

(1) Is not the concept of context-building systematically ambiguous, depending upon whether we are interested in looking at science and scientific change from outside or we are interested in looking at them from inside?

(2) What implications follow from the Heisenberg-Einstein context principle, reconnecting observation with theory in a manner which is quite at variance not just with their own earlier narrow convictions or beliefs but with those of the philosophers including the logical positivists?

We think that it has become more important today than ever before to distinguish clearly science-in-context - what the philosophers called the context of discovery - from the context in science. The philosoher-methodologists may choose to deal, or not to deal, with the former. But they cannot ignore the latter. In particular, they cannot ignore the context in physics and in other natural sciences. In order to differentiate it rather sharply from the philosophical conception of the context of discovery in the twentieth century tradition of Carnap, Reichenbach and Popper, we have proposed to designate it in this book as the core-context of theory development. Once clearly recognized, the idea of the core-context of theory development implies that context-building within physics plays a dynamical role as the current theories mature sufficiently enough, not only creating new frontiers of research but demanding unification of the fundamental laws of nature. Quantum mechanics provides a good example of the context-building in physics which we are here hinting at.

More generally, in order to answer the question (1) above in the affirmative, we want strongly to argue that the dominant idea of science-in-context cannot be allowed to blind us to the important role which the context principles play inside the natural sciences. In particular, it cannot be allowed to blind us to the dynamic processes of context-building inside physics, constraining the development of physical theory as physics makes progress from one frontier to the other. But that is exactly what has happened during the past century, thanks to the logical positivists generally, and to famous philosophers like Reichenbach, Carnap and Popper particularly.

Historically, the French physicist-philosopher, Pierre Duhem (1906, 1914, 1954) was the first to hint at a context principle, which is relevant to the rationality of theory-testing in physics, by arguing as follows:

> "the physicist can never subject an isolated hypothesis to experimental test, but only a whole group of hypotheses; when the experiment is in disagreement with his predictions, what he learns is that at least one of the hypotheses constituting this group is unacceptable, and ought to be modified; but the experiment does not designate which one should be changed (Duhem 1954, p. 187)."

To bring out the full significance of Duhem's holism (Pandit 1982, p. 112), as it is popularly known among the twentieth century philosophers, it is necessary to re-state it along the following lines (Swanson 1967, p. 59):

> (1) a single scientific hypothesis H is never disconfirmable in isolation from its fellows; (2) every single hypothesis H of science presupposes, explicitly or implicitly, the support of a conjunction $A = A1 \cdot A2 \cdot Am$ of auxiliary assumptions or hypotheses; (3) the failure of an observational consequence of H in the face of contradictory evidence disconfirms only the conjunction of H and A, not H alone - i.e., establishes only $\sim (H \cdot A)$.

Thus, Duhem is mainly concerned with theory-testing in physics, where clusters of theories, or hypotheses, participate while a newly proposed theory undergoes test. Given the theory under test, the test-results are to be attributed to all the theories taken as a whole, which participate in the test. If the test-results are negative because the predictions of the individual theory in question fail, no single theory in the cluster can be held responsible. Theory-choice would also be affected. On the way to interpreting the implications of Duhem's methodological holism, the twentieth century analytic philosophers of science invented a thesis, the so-called 'Duhem-Quine thesis' (Lakatos 1970, 180-189), which says: " given sufficient imagination, any theory can be permanently saved from 'refutation' by some suitable adjustment in the background knowledge in which it is embedded."[33] This clearly shows that no attention has been paid to the possible role context principles can play in physics. In the discussions on Duhem, the attention shifted instead to the *ad hoc* hypotheses and to 'conventionalist stratagems' (Popper 1934, sections 19 and 20, Lakatos 1970, pp. 117-119).

Unlike the questions most favoured by the dominant traditions of the twentieth century philosophy of science, we would like to emphasize here how important it is to the methodology of science to raise the following questions:

1. Whether there build up core-contexts of theory development within physics?

2. Whether such contexts build up where theories, which are already successful and confirmed, participate in the process of discovering the kind of new theory which physics can aim at?

We think that there are at least three kinds of possibilities which we must consider. First, in physics a core-context of theory development emerges when the existing theories are not only successful but acquire sufficient maturity to become frontier-creating theories. Take, for example, Einstein's general theory of relativity which received experimental confirmations already in 1919. After remaining out of focus, it came back on the scene in 1960s, playing a fundamental role at the new frontier now known as the Standard Model of cosmology. Of course, the participating theories or theory must also be successful in other important respects. The general theory of relativity amply fulfils this requirement.

Secondly, a core-context of theory development may emerge on the horizon when the participating theories are those which provide crucial inputs to the methodology of unification in physics. They must all

---

[33] A detailed discussion can be found in Lakatos (1970, pp. 91-196).

participate in the search for a new theory in which the earlier theories would be allowed to live contradiction free. For example, this is true of quantum mechanics and the general theory of relativity. At present they cannot live together contradiction free. The search for a quantum theory of gravity is on in fundamental physics. If a theory is found which unifies all the four forces of nature, it may be possible for the existing theories to live in it contradiction free.

Thirdly, if we reconnect observation with theory in accordance with the Heisenberg-Einstein context principle, then physics can aim at those theories which fulfill this condition rather than those which do not. Thus, to think of the core-context of theory development in physics is to think of those theories about which we can say at least three things: (1) that they not only explain all that which they are expected to explain but they decide what can be observed or measured; (2) that, being highly successful, they participate in the methodology of unification; and (3) that, being sufficiently rich, mature and powerful, they create new frontiers of discovery, making physics open-ended.

Taken in these three dimensions, the core-context of theory development in physics deserves serious methodological attention. We think that the future of the methodology and philosophy of science will depend upon how we go about the task of understanding the core-context of theory development in physics and, possibly, in other sciences. Very modestly speaking, while trying to place Heisenberg-Einstein dialogue on physical theory within a broader framework, our aim here is to take another look at their dialogue which had a beginning in early 1926 in Berlin, when Einstein himself, after listening to his lecture on quantum mechanics, took the initiative for opening a discussion with young Heisenberg (1984, Band II, 1956-1968, S. 428). If we want to know who it was exactly who took the first look at their dialogue, we will find that it was Heisenberg himself, who has left us an account while doing so, probably the only account which deserves serious study.

There arise several important questions concerning their dialogue. The first question is which particular aspect of physical theory - or rather which aspect of theory-finding, theory-testing and theory-choice - were Heisenberg and Einstein in their dialogue concerned with. The second question is whether there emerged from their dialogue some context principle, as a guiding principle of sufficient methodological interest, on which they both showed significant agreement, resolving some of their methodological differences by removing some of their shared philosophical misunderstandings concerning physical theory. And the third question is whether their dialogue helped them in arriving at a better understanding of quantum mechanics, resulting in a significant impact on their respective positions on its foundations. Whether the principle which emerged from their dialogue could be placed in a wider methodological framework within physics itself? All of these questions are methodological questions. As regards the second and third questions, which are rather closely linked with each other, we shall argue that they can be answered in the affirmative. In particular, as regards the third question, the impact in the case of Heisenberg appears, at first sight, to have been far-reaching, or rather far more dramatic than in the case of Einstein. But, in the final analysis, in his case it must be seen in the light of the mathematical formalism of quantum mechanics including the uncertainty relations, while in the case of Einstein one must look at it in the light of the developments that had already accompanied the origin, the formulation and the experimental confirmation of his general theory of relativity. In particular, we must look at it in the light of Einstein's own context principle (Einstein 1919, S. 52; and Einstein1960, p. 77)) which says:

> "Es ist das schönste Los einer physikalischen Theorie, wenn sie selbst zur Aufstellung einer umfassenden Theorie den Weg weist, in welcher sie als Grenzfall weiterlebt."

As a result of their dialogue, their earlier methodological positions underwent a significant change, with both of them embracing the same guiding principle (Heisenberg 1979, S. 13):

> „Erst die Theorie entscheidet darüber, was man beobachten kann."

Rendered into Einglish, it says that it is always the theory which decides what can be observed (Heisenberg 1979, S. 39). On the other hand, in our response to the first question, we shall frequently refer to this very principle as the Heisenberg-Einstein context principle, their most important methodological meeting point, which must have even shaped their later maturer understanding of the foundations of quantum mechanics. Thus, while focusing on Heisenberg-Einstein context principle, we shall argue that both of them were deeply concerned with those methodological aspects of physical theory which dominate the context-building both in theory-finding and theory-testing, generating new frontiers of discovery in physics and astronomy. Any change of theory, or its further development, can take place within relevant contexts that build themselves up within physics. But this has nothing to do with that which the leading philosophers of physics of the 20th century called the context of discovery, by which they meant science in context, distinguishing it from the context of justification of complete, possibly rival, theories. In a nutshell, our present project is to understand how the Heisenberg-Einstein context principle, and the entire dialogue on which it is based, contribute to our understanding of the context-building in physics, reconnecting observation and measurement with theory.

In order to put forward here our main claims, we shall proceed on the following assumption while relating it to the Heisenberg-Einstein context principle. From time to time, there develop from the very core of a physical theory - from its mathematical formalism - the contexts which can function as the dynamic core-contexts of theory development, taking it to the very frontier of discovery - die vorderste Front der Forschung. As we have seen, our choice of this designation for such specific contexts derives from several considerations, the most important among them being the methodology of working forwards from physical theories to those problems whose solutions are not yet in sight (Pandit 1982). In a nutshell, this is the methodology (Pandit 1982, 1991) which requires that a good physical theory T must possess resolving power - $T_{RP}(T)$ - besides its explanatory power - $T_{EP}(T)$. The world which the $T_{EP}(T)$ is by its very design intended to explain can be, for good reasons, approached as an object of scientific interrogation. Nature explained can at the same time continue to be nature interrogated as to its unknown and undiscovered aspects. The specific contexts which develop within physics for this purpose are determined by the $T_{RP}(T)$ and not by any of those factors that might be external to the scientific activity of world-making by explanatory theory-forming, theory-testing and theory-choice. As is quite possible, our enterprise here may sound not only ambitious but rather unconventional - one which is far removed from the mainstream philosophy of science. Should this be the case, we will be more than contented by staying away from all that which can be described as the failure of the 20th century philosophy of science. We are here referring to the approach to science chosen from outside science, which led the eminent physicist Steven Weinberg (1992) to devote a whole chapter in his book *The Dreams of a Final Theory* to the theme "Against Philosophy", contrasting the "unreasonable effectiveness of mathematics" with the "unreasonable ineffectiveness of philosophy" in the development of physics. The practice of the 20th century philosopher of science can be compared with the practice of someone whose beliefs do not fit into the standard practice of the physicists and astrophysicists, as if one could do astrophysics by looking at stars as finished products, and not by looking at star-formation, star-birth, star-explosion, supernova-explosion and so on. In any case, we intend here to restrict ourselves to an enterprise which was already being hinted at as early as 1916 by Albert Einstein when he said: "Es ist das schönste Los einer physikalischen Theorie, wenn sie selbst zur Aufstellung einer umfassenden Theorie den Weg weist, in welcher sie als Grenzfall weiterlebt" ((Einstein 1919: 52, 1960: 77). If a physical theory is at all to play this role in physics and astronomy, then it is most appropriate to study the dynamic core-context building that creates new frontiers of physical discovery.

The problem of understanding the role which inter-theoretical relations play in context-building in physics and in creating new frontiers of discovery presents itself as a challenging task for the methodology and philosophy of physics. In order to have a closer look at the core-context building in current fundamental physical theory, let us ask, with Eugene Wigner (1982, 124), 'How close are our present theories to perfection? ... are they, in particular our physics, even self-consistent? The answer to this question is rarely publicized, but it surely is: no. The general theory of relativity is based on the assumption of the meaningfulness of the space-time point concept defined by the crossing of two-world-lines, yet it is easy to show that quantum mechanics does not permit the definition of such points. The basic idea of the fundamental interpretation of quantum mechanics postulates the process of "measurement". Yet it is easy to show that the existence of this process is not consistent with the principles of quantum mechanics'. Wigner (1982, 125) argues further: "Quantum mechanics is not in complete harmony with the theories of relativity, particularly not the general one. And even quantum mechanics alone still lacks the complete simplicity which we are striving for - in spite of the accomplishments of the past, particularly those of Salam and Weinberg, we still have several types of interactions, not united into a single equation, and there are other grave problems". His following observations on the developments within physics are rather directly concerned with the subject of inter-theoretic relations: "The development of science, in particular physics, is miraculous for another reason: every step in its development shows that the preceding theory was valid only approximately, and valid approximately only under certain conditions. Newton's theory is valid with a high accuracy if only gravitational forces play a role, all pre-quantum theories are valid only for macroscopic bodies and for these only under certain conditions. It is lucky that such special conditions exist under which simplified approximate theories present a wonderfully good approximation. Surely, if macroscopic theories had not been developed, it would have been even more difficult, perhaps impossible, to develop quantum mechanics. General relativity would not have been invented, not even by Einstein, had the original theory of gravitation not existed" (Wigner 1982, 124).

What are then the most important, or at least distinguishing, features of contemporary physics at the frontier of discovery? Does it in any way indicate that revolutions in physics are driven by observations and experiments alone? Can we not situate them in the context of a set of core-theories and core-problems which, together with the experimental situation, could define physics at the frontier of discovery? What is it which decides what kind of fundamental theory can physics at the frontier of discovery aim at? We think that at any particular stage of scientific progress the context in which revolutions in physics take place can be identified with the set of core-theories and core-problems, making physics open-ended at the frontier of discovery. Taking these as constituting the theory-problem interactive systems (Pandit 1982, 1991), we may designate them as dynamic core-contexts of development in physics. At the end of the 19th century, physicists generally looked at (Newtonian) physics as complete, thinking that the task that still remained was a matter of filling in detail only. But at the beginning of the 20th century, it became clear that the mature theories of Maxwell and Newton played rather the role of the core-context of development for many new developments in physics, notably Einstein's special and general theory of relativity and the quantum theory. Again, think of Maxwell's unification of electric and magnetic forces in 1860's, predicting that light is an electromagnetic phenomenon. In the 1920s, Maxwell's theory and Einstein's general theory of relativity, with all its novel predictions and their confirmations, formed the core-context of theory development for attempts at further unifications, including Einstein's own unsuccessful search for a unified field theory. On the other hand, it is a well-known fact that the core-problem of blackbody radiation shaped physics at the frontier of discovery from 1859 until 1926, not only in thermodynamics, but in electro-magnetism, in the old quantum theory, and in quantum statistics, in that order (Pais 1982, 28). Thus, in the kind of development of fundamental physical theory which physics at the frontier of discovery can aim at, what is most important is how the core-context of theory development builds itself up in theory-testing and theory-finding.

How strongly do theory-testing and context-building interplay with each other in shaping the frontier of discovery in physics? While theory-testing by checking a candidate theory by setting up novel experiments can result either in its confirmation or in its falsification, context-building can develop the core-context of theory development from the very core of the existing theory, or theories, indicating the path to formulating the new theory which physics aims at. The stronger and the richer the confirmations resulting from theory-testing, the greater will be the maturity of the theory under test. The more mature a theory becomes by testing its consequences, the greater will be its ability to develop into a frontier creating core-context of theory development. Today, at the end of the 20th century, quantum mechanics itself serves as the most mature physical theory in this sense, where experimental confirmations of the theory are concerned. With Heisenberg, we can say that the quantum core-context building has already taken place. Now the most important task is to explore how far it can be developed further into a dynamic core-context of theory development.

Heisenberg's uncertainty principle states that the values of both the members of certain pairs of canonically conjugate variables, such as position and momentum, cannot be determined simultaneously to arbitrary precision. The uncertainty relations, as they are called,

$$\Delta q \; . \; \Delta p \geq \bar{h}/2$$

express mathematically this irreducible level of uncertainty of the relevant pairs of dynamical variables when they are measured together. Thus, there is a reciprocal relationship between them, such that the product of their uncertainties must be greater than Planck's constant *h.* A universal statement of prohibition at the quantum level, this has to be understood as an important consequence of the formalism of quantum mechanics itself. As an intermediate step in the development of fundamental physical theory, consider Dirac's unification of quantum mechanics with the special theory of relativity and electrodynamics in the 1920s, resulting in the relativistic quantum mechanics of the electron. This belongs rather to the period of development of physical theory when most physicists were preoccupied with the fundamental difficulties of understanding the stability of atoms, the way electrons interact within an atom and the non-classical properties like the spin of a sub-atomic particle. Among other physicists, including Schrödinger, who clearly recognized the problem-situation developing within physics, it was Dirac who succeeded in finding a solution to the core-contextual problem of the existence of the negative energy states for the electron. It is necessary to ask what kind of context-building was at work here. For a physicist, like Dirac, wanting to write down a complete equation for the electron, there was a context-building from the discovery of the quantized angular momentum of the electron by O. Stern and W. Gerlach in 1922 and from the very core(s) of the special theory of relativity on the one hand and quantum mechanics on the other. Dirac knew that anyone grappling with the problem of the existence of negative energy states would have to work within this very context. The kind of theory-testing and context-building, which we are here referring to, got accelerated with the development of quantum mechanics, once it was already known since 1905 that "Einstein's formula for the energy of a system with a given momentum involves a square root, and the result is that the value for the energy, mathematically, can be either positive or negative" (Dirac 1983, 51). With such core-context-building, it was no longer possible to disregard the problem of the negative energy states. Quantum mechanics allows jumps discontinuously from one energy level to another. In the words of Dirac (1983, 50), "If we start off a particle in a positive energy state, it may jump into a negative energy state." If we cannot exclude the negative energy states from our theory, Dirac (1978, 16) thought, we must find a method of physical interpretation for them. It was in May 1931 that Dirac (1978, 16) proposed that there exists a new kind of particle ' "unknown to experimental physics, having the same mass and opposite charge to an electron. We may call such a particle an anti-electron".' Published in the Proceedings of the Royal Society in

September of that year, it announced the birth of the modern idea of antimatter (Fraser 2000, 62). The far-reaching consequences of his theory's novel prediction that anti-electrons exist was confirmed in 1932 by the experimental discovery of the positron by C. D. Anderson.[34] Today we can look back to Dirac's relativistic quantum theory of the electron as a theory inseparable from context-building in physics at the frontier of discovery where the high energy collision experiments with particle creation (transmutation of energy into matter in accordance with the special theory of relativity) play quite a dominant role. Here it would be quite wrong to think of the concept of a fundamental field - or rather of a fundamental elementary particle - as if it was no longer well-defined simply because particles created by particle collisions, e.g., in electron-positron collusion experiments, have only a virtual existence. On the contrary, in fundamental physical theory, symmetries and fundamental fields are equally important.

We consider it an important task of the physicist and the methodologist-philosopher of physics to find out how, from time to time, there develop within physics the core-contexts of theory development in which the physicist is constrained to make his/her choice between the possible paths for finding a new theory. Thus, here we are not at all concerned with physics in context. Our sole concern is rather the context within physics. Normally, a physical theory itself matures under experimental testing until it itself becomes a dynamic core-context of theory development, raising new problems and generating new frontiers. Dynamic core-contexts of development can build themselves up within physics even from the core of those fundamental theories which have, after acquiring maturity, become controversial either because of their mutual consistency being in question or the very path to their unification being a highly problematic and thorny one. For example, consider the fact that the general theory of relativity has still not been successfully incorporated into a consistent quantum mechanics. The physicists agree that quantum gravity is one of the great frontiers of discovery at the beginning of the 21th century. There may be alternative possibilities of approaching the current problem of formulating a quantum theory of gravity. But in this kind of situation it is quite important to ask: What is the dynamic core-context of development for finding the path that leads to such a theory? What

---

[34]The similar anti-particle on the nuclear level, the antiproton, was discovered by Chamberlain, Segre, Wiegand, and Ypsilantis in 1955.

are the core-problems which contribute to the context-building for the physicist's search for such a theory? If we only think of the problem of understanding how the fundamental particles acquire their masses (the problem of the origin of mass),[35] or of the problem of the cosmological constant as the most serious of all problems in particle physics and cosmology, or of the problem of understanding the beginning of the universe and its present rate of expansion (including the dark matter problem - why does most of the matter that gravitates in the universe seem to be invisible?), a quantum theory of gravity seems to be the kind of theory which physics at the frontier of discovery can aim at.

The current physical theories always build up an implicate order of inter-theoretic clustering, partly harmonious, partly full of tension. The inter-theoretical relations which they build up trigger the search for a new theory T not yet in sight, while restricting, at the same time, the choices available to physics at the frontier of discovery. Over historical time, they develop within physics what we have chosen to designate as dynamic core-contexts of theory development, which enable it then to find its path to T. We think that the best examples of dynamic core-contexts of theory development come from the history of the methodology of unification in physics. The methodology works by building up enough symmetry into the fundamental laws of nature as and when these are formulated, so that the underlying fundamental interactions and their force laws are unified, resulting in their description by a single unified theory. Interestingly enough, this methodology has itself evolved over a period of time and acquired considerable complexity, possibly with a feedback from fundamental physical theory itself. For example, the 1960s saw a kind of shifting away from the earlier insistence on a reductionist research programme of unification by an ultimate explanation of nature in terms of a final theory of the simplest elementary particles, rather in the old tradition of Democritus. But the programme staged a comeback in the 1970s mainly through the development of the electroweak theory and the QCD, with the emphasis shifting to the local quantum field theories

---

[35]The Large Hadron Collider (LHC), located at CERN, Geneva, Switzerland, the world's largest and highest energy and highest intensity particle accelerator is the best and the latest example not only in technological advances but in the dynamic interface between theory and experiment in physics. The Search-and-Discovery-procedures in this context have as their target the predicted Higgs boson, the electrically neutral spinless particle which is the most elusive object in the Standard Model of particle physics. It is currently believed that the high energy and precision of an electron-positron linear Collider holds an important key to the next step in a comprehensive understanding of the laws of particle physics.

in which both the fundamental fields - or their particle representations - and fundamental symmetries interplay rather intimately. And this is the present scenario in particle physics. As physics has moved from one frontier to another in search of the unifying theory, Tu, its dynamic core-contexts of theory development have exhibited a fundamental continuity in terms of the mathematical formalisms, from one successful unification to the other. On the other hand, they also have exhibited a strong undercurrent of continuity in the following equally non-trivial sense: As they have created new frontiers of development, the older problems have been seen in a new light, as also the older theories which had themselves either posed those problems or provided their solutions. This type of continuity can be best described in terms of the earlier stable forms of knowledge being reproblematized by the dynamic core-contexts of theory development.

Consider in this context the remarkable success of physics in unifying the electromagnetic force with the weak force, making it possible to look beyond the electroweak theory and Quantum Chromodynamics (QCD) to the grand unified theories (GUT). The current Standard Model of the elementary particle physics, a gauge theory of the strong and electroweak interactions, provides a fundamental theory of quarks and leptons, which has been tested up to energies approaching 1000 GeV. The important question about particle physics at very short distances, which pre-occupies the physicists at the frontier of discovery, is how one might go further from the Standard Model to its possible extension to supersymmetry, or superstring theory, or grand unification, making a more comprehensive understanding of the laws of particle physics and, therefore, of the earliest history of the universe possible.

One might ask what is it which makes physics an empirical science. Those who look at it from outside generally believe that, in physics, "concepts and mathematical constructs can simply be taken from experience" i.e., from empirically established data (Heisenberg 1985, 444). Did either Einstein or Heisenberg hold this kind of view at any time during their life-time? Heisenberg (1979, 13) had himself once wondered:

> "If this was the whole truth, when entering into a new field, we should introduce only such quantities that can directly be observed and formulate natural laws only by means of these quantities."

As a young man, recalls Heisenberg (1979, 13), he also believed "that this was just the philosophy which Einstein followed in his theory of relativity." He is here referring to Einstein's special theory of relativity. What is more interesting is that Heisenberg first thought, quite consistent with his belief, as if he too was following the same kind of philosophy - the philosophy of introducing only observable quantities in one's theory. Sooner or later one must ask how far such a requirement is appropriate for a physical theory. How far is it possible for a theory to be constructed solely in terms of quantities which can directly be observed, or which are closely related to such observable quantities? If you put the theory you construct in chains and expect it to save the phenomena, it reminds us more of Ernst Mach's overly restrictive phenomenalistic empiricism, rejecting atomism in physics. It even reminds us of logical positivism of the Wiener Kreis, demanding, in the tradition of the classical British empiricism and Ernst Mach's phenomenalistic empiricism, that a physical theory be verifiable in principle, as if the methodological rules developing from within the twentieth century physics were of no interest to us. It is remarkable that today we can say with certainty how ineffective the postulate of observability has been as it did not play any significant role in context-building in the major developments in physics during the past century. Had it done so, the development of physical theory would have been crippled rather from the very early stage. But the important question still remains:

> Did Einstein and Heisenberg have to change their views
> on the aim and structure of physical theory and on its
> connection to observation, or measurement?

The answer is yes. We shall consider below how this change in their approaches came about. The biggest drawback of the strategy of raising the observability of all quantities, e.g., energy, frequency, and so on, in one's equations to a postulate, as the young Heisenberg and, before him, Einstein and others did under the influence of the Mach-type phenomenalistic empiricism, is that it encourages one to do physics in a style as if saving the phenomena were the sole aim of a physical theory.

Since both of them rejected later on the view that the postulate of observability had played a role in the construction of physical theory, what were the reasons, possibly developing from within physics itself, which led to such a development? To find an answer, we have to go back to early 1926, when under the auspices of the Physics Kolloquium of the University of Berlin, and in the tradition of Hermann von Helmholtz, Heisenberg was invited to report on the subject of the newly developed quantum mechanics. After the Kolloquium, which was attended by Einstein, among others, he and Heisenberg met at former's Berlin flat where an interesting discussion followed. Quite naturally, the discussion was started by Einstein with a question as follows: "What was the philosophy underlying your kind of very strange theory? The theory looks quite nice, but what did you mean by only observable quantities?"[36] Somewhat more specifically, he asked him

> Was Sie uns da erz̈ahlt haben, klingt ja sehr ungeẅohnlich. Sie nehmen an, daß es Elektronen im Atom gibt, und darin werden Sie sicher recht haben. Aber die Bahnen der Elektronen im Atom, die wollen Sie ganz abschaffen, obwohl man doch die Bahnen der Elektronen in einer Nebelkammer unmittelbar sehen kann. Können Sie mir die Gründe für diese merkwürdigen Annahmen etwas genauer erklären? [37]

Recalling his own reply to this question, Heisenberg (1984, Band II, 1965-1968, S. 428) admits that, "I did not believe any more in electronic orbits, in spite of the tracks in a cloud chamber. I felt that one should go back to those quantities which really can be observed and I also felt that this was just the kind of philosophy which he had used in relativity; because he also had abandoned absolute time and introduced only the time of the special coordinate system and so on". To this Einstein had responded by saying "That even if this may have been his philosophy and even if he may have used it, it was nonsense all the same, which meant that it was completely wrong (Heisenberg 1985, Band III, 1969-1976, 459-460)." The most significant remark which Einstein then made was that "It is never possible to introduce only observable quantities in a theory. It is the theory which decides what can be observed (Heisenberg 1985, Band III, 1969-1976, 459-460)."

---

[36]Heisenberg (1984, Band II, 1965-1968, S. 428). Here Heisenberg remarks: "But when I had to give a talk about quantum mechanics in Berlin in 1926, Einstein listened to the talk and corrected this view."

[37]Werner Heisenberg (1979, S. 30). Here Heisenberg (S. 30) recalls Einstein's deeper concern when the latter asked him to clarify: "Aber Sie glauben doch nicht im Ernst, daß man in eine physikalische Theorie nur beobachtbare Größen aufnehmen kann." Also turn to S. 91.

In the epistemology and methodology of the physicist's search for physical theory, if we want to single out at least one methodological principle of foundational importance to physics which has dominated the discussions between Heisenberg and Einstein, it is the one which Heisenberg has himself frequently focused on in his essays and lectures, almost invariably attributing it to Einstein during their intimate discussions on quantum mechanics, particularly on Heisenberg's uncertainty relations. The main aim of their discussions was to resolve their differences in order to improve their respective positions concerning the interpretation of quantum mechanics. The methodological principle (or the Heisenberg-Einstein context principle) as cited above, which they both came to agree upon, says:

> "Erst die Theorie entscheidet darüber, was man beobachten kann ..."

In the case of Heisenberg, one notices in the light of this principle that there is an interesting, a non-trivially significant, turning around of the fundamental questions in quantum mechanics with far-reaching consequences for our understanding of its interpretational problems. What makes it all the more significant is Heisenberg's own acknowledgement as follows: In his many lectures and essays, he (Heisenberg 1979, S. 13) returns to this theme while clarifying the nature of the development of concepts in the physics of the twentieth century, particularly in the fields of quantum mechanics and elementary particle physics.

In our view, Heisenberg's account of his dialogue with Einstein deserves a serious attention, at least for the following reasons. First, we think that it carries within itself a great methodological insight, provided we are interested in understanding what exactly the Heisenberg-Einstein context principle, as a guiding principle, says or implies in the context of the physicist's search for fundamental physical theory. If it is the theory which decides what can be observed, or measured, then it is methodologically imperative for the physicist working at the frontier of discovery to ask: (i) What are the theories in the background which serve as, or provide, the core-context of development; and (ii) what kind of fundamental theory can physics aim at? Thus, with the help of this principle, one can explore important correlations that may obtain between the aim and structure of a physical theory, or between the frontier of discovery and the core-context of development. Secondly, Heisenberg's (attitude to his) own discovery of the uncertainty relations, no less than the discoveries made earlier by Einstein, demonstrates the methodological importance of this principle as a guiding principle in physics. Thirdly, as regards Einstein himself, given this principle, the deeper epistemological and methodological reasons for his view that quantum mechanics is incomplete become rather clearly visible and accessible to rational reconstruction along the following lines of argument: Even before the physicists decide to legitimately subject the extant theories to severe criticism by experimental test, and by serious epistemic appraisal, they had better decide first what kind of fundamental theory should physics aim at. Given the kind of theory which physics should aim at, they can then decide what kind of methodology, not only of appraising but of finding a theory, should it develop and adopt. Both these steps seem essential to the building up of the core-context of development within physics for a better understanding of the extant theories and the role they themselves can play for seeking better alternatives to them, where possible or necessary. Thus, the very rationality of Einstein's radical opposition to (Copenhagen interpretation of) quantum mechanics, and to the general acceptance of the theory as if it were a final theory, could be explained with the help of this principle, whether or not it is taken together with the Bohr-Einstein debate. Fourthly, in our opinion, no rational reconstruction of Einstein's own research programme would be complete, if it did not find a fundamental role for this principle. Einstein's belief in this principle seems to have been so firm, as if it was that part of the Galilean methodology which a physicist could not easily give up. Fifthly, the real, but largely unacknowledged, impact of the Heisenberg-Einstein context principle on the developments in the twentieth century physics and philosophy have yet to be properly assessed. Sixthly, and lastly, one might ask the question:

> What was the exact nature of the impact of Heisenberg's Berlin lecture on Einstein? How far-reaching were the consequences of their subsequent discussion for Einstein himself and for Heisenberg, or for Bohr? Did their views stand corrected in some sense by this methodological principle which originally emerged from Einstein's remark?

If we follow the details of their dialogue as reconstructed by Heisenberg, clearly the answer is that it made a most significant contribution by making an improved understanding of quantum mechanics possible, where Einstein, Heisenberg and Bohr were concerned. In Heisenberg's (1984, Band II, 1956-1968, S. 429) own words:

> "This remark of Einstein was very important for me later on when Bohr and I tried to discuss the interpretation of quantum theory ..."

The task of clarifying its nature and implications assumed a great urgency for Heisenberg (1985, III, 1969-1976, S. 446), since "The most conspicuous demonstration of this thesis by Einstein was the relations of uncertainty". The question of the nature of the methodological principle expressed by Einstein's remark assumes importance when considered in the context of Heisenberg's acceptance of its role in the interpretation of quantum mechanics. This question will be taken up for discussion in what follows.

Before considering how the Heisenberg-Einstein context principle, as it emerged from their dialogue, led to a change of view and to a change in attitude - to an attitude of turning those questions around which one usually asked those days in physics, especially in the context of quantum mechanics - let us pay some attention to Einstein's argument in support of his view that "whether you can observe a thing or not depends on the theory which you use. It is the theory which decides what can be observed" (turn to Heisenberg 1984, Band II, 1956-1968, S. 429). As reconstructed by Heisenberg (1984, Band II, 1956-1968, S. 429), the argument runs as follows:

> "Observation means that we construct some connection between a phenomenon and our realization of the phenomenon. There is something happening in the atom, the light is emitted, the light hits the photographic plate, we see the photographic plate and so on and so on. In this whole course of events between the atom and your eye and your consciousness you must assume that everything works as in the old physics. If you would change the theory concerning this sequence of events then of course the observation would be altered".

Did Einstein's view that it is really dangerous in physics to say that one should only speak about observable quantities, a view which

Heisenberg also now shared, have damaging consequences for those methodologies - recall Ernst Mach's attitude to the concept of atom - which are notorious not just for their instrumentalistic character but for their ineffectiveness in the development of science? The answer is clearly in the affirmative. Given the postulate of observability concerning how to structure a physical theory, what one could expect a theory to do was just to save the phenomena. Moreover, if Bohr and Heisenberg had held the view that the equations of quantum mechanics are a mathematical tool of calculating the probabilities for the various outcomes of measurements (Heisenberg 1984, Band II, 1956-1968, S. 429), this did not necessarily imply that they were favoring instrumentalism in a rather philosophical sense, as some philosophers have suggested by way of criticism. We think that their view is not incompatible with Heisenberg-Einstein context principle which asserts that it is always the theory which decides what can be observed. On the contrary, it can very well serve as a demonstration of the principle when extended to the context of the interpretation of quantum mechanics, far beyond the domain of classical objects. According to Heisenberg (1984, Band II, 1956-1968, S. 429):[38]

> "In quantum theory it meant, for instance, that when you have quantum mechanics then you cannot only observe frequencies and amplitudes, but for instance, also probability amplitudes, probability waves and so on, and these, of course, are quite different objects."

But when a change of theory results in change in observations, what happens to the pre-existing concepts of a pre-existing theory? This is precisely the kind of question which Heisenberg himself (1984, Band II, 1956-1968, S. 429) asks as follows: "... when one has invented a new scheme which concerns certain observable quantities, then of course, the decisive question is: which of the old concepts can you

---

[38]In an interesting statement regarding the role of the state vector, Eugene Paul Wigner (1971: 5-6) says: "There are two epistemological attitudes toward this. The first attitude considers the state vector to represent reality, the second attitude regards it to be a mathematical tool to be used to calculate the probabilities for various possible outcomes of observations. It is not easy to give an operational meaning to the difference of opinion which is involved because, fundamentally, the realities of objects and concepts are ill defined. One can adopt the compromise attitude according to which there is a reality to objects but quantum mechanics is not concerned therewith. It only furnishes the probabilities for the various possible outcomes of observations or measurements - in quantum mechanics these two words are used synonymously." See Eugene Paul Wigner (1971) "The Subject of Our Discussions", in B. d'Espagnat (ed.) Foundations of Quantum Mechanics, International School of Physics "Enrico Fermi" 1970. Academic Press, New York, 1971, pp. 1-9.

really abandon? In the case of quantum theory it was more or less clear that you could abandon the idea of an electronic orbit."

According to Heisenberg (1984, Band II, 1956-1968, S. 429), what Einstein must have meant by his remark is "that when we go from the immediate observation - a black line on a photographic plate or a discharge in a counter - to the phenomena we are interested in, we must make use of theory and of theoretical concepts. We cannot separate the empirical process of observation from the mathematical construct and its concepts". Interestingly enough, Heisenberg's (1985, Band III, 1969-1976, S. 459-460) own reformulation of this idea runs as follows: "In order to understand Nature we have to approach it by some concepts; and we try to establish an immediate connection between the observed phenomena and the concepts. If we are successful, we have both defined what we have observed and confirmed the validity of the concepts; if not, we may be forced to change the conceptual frame."

The question which one might ask here is whether an alternative way of looking at the Heisenberg-Einstein context principle is possible, which might bring us closer to understanding quantum mechanics, the uncertainty relations in particular, Einstein's attitude to it and the attitude of Heisenberg himself. In order to answer this question, we must go back to early 1926 when the interpretational problems of quantum mechanics were still tormenting Heisenberg in Copenhagen. "...we felt", to quote Heisenberg (1985, Band III, 1969-1976, S. 453), "that in the atom it seemed all right to abandon the concept of an electronic orbit. But what in a cloud chamber? In a cloud chamber you see the electron moving along the track; is this an electronic orbit or not?" How many nights must have he and Bohr spent discussing these problems, as Heisenberg (1984, Band II, 1956-1968, S. 433) recalls, with Bohr emphasizing the dominant role of the wave-particle dualism while he himself thought of the mathematical formalism as his starting point in search of a consistent interpretation. It was around this time that he (Heisenberg 1984, Band II, 1956-1968, S. 433) remembered Einstein's remark that "It is the theory which decides what can be observed". There was, as a result, such a change in the whole approach, such a change in the very understanding of the problems that everything fell in place. Let us quote Heisenberg (1984, Band II, 1956-1968, S. 433) as he describes this change in the following words:

'From there it was easy to turn around our question and
not to ask: "How can I represent in quantum mechanics

this orbit of an electron in a cloud chamber?" but rather to ask "Is it not true that always only such situations occur in nature, even in a cloud chamber, which can be described by the mathematical formalism of quantum mechanics?" By turning around I had to investigate what can be described in this formalism; and then it was very easily seen, especially when one used the new mathematical discoveries of Dirac and Jordan about transformation theory, that one could not describe at the same time the exact position and the exact velocity of an electron; one had these uncertainty relations. In this way things became clear. When Bohr returned to Copenhagen, he had found an equivalent interpretation with his concept of complementarity, so finally we all agreed that now we had understood quantum theory.'

Turning around the questions which one usually asked, Heisenberg (1984, Band II, 1956-1968, S. 433) posed the question: "Well, if we want to know of a wave packet both its velocity and its position what is the best accuracy we can obtain, starting from the principle that only such situations are found in nature which can be represented in the mathematical scheme of quantum mechanics?" Posing such type of questions resulted in the principle of uncertainty, which seemed to be compatible with the experimental situation - e.g., the fact that "the path of an electron in a cloud chamber was not an infinitely thin line with well-defined positions and velocities" (Heisenberg 1985, Band III, 1969-1976, S. 410-411). According to this principle, "the wave packet representing the electron is changed at every point of observation, that is at every water droplet in the cloud chamber. At every point we get new information about the state of the electron; therefore we have to replace the original wave packet by a new one, representing this new information (Heisenberg 1985, Band III, 1969-1976, S. 410-411)."

Let us now briefly consider where this kind of theory and this kind of questioning can lead us. Some of the consequences are described by Heisenberg (1985, Band III, 1969-1976, S. 411) as follows: "The state of the electron thus represented does not allow us to ascribe to the electron in its orbit definite properties like coordinates, momentum and so on. What we can do is only to speak about the probability to find, under suitable experimental conditions, the electron at a certain point or to find a certain value for its velocity. So finally we have come to a definition of state which is much more abstract than the original electronic orbit. Mathematically we describe it by a vector in Hilbert space." The Hilbert space being a space of infinitely many dimensions, the concept of state here diverges from that in classical physics. The divergence does not yet mean any major departure from the methodology of physics as formulated by Galileo. We can still look at nature as the object of physical knowledge. Nature can still be described, or known, objectively by means of the mathematically formulated theories.

The most important question which has remained unasked in the context of the interpretational problems of quantum mechanics is how far did the Heisenberg-Einstein context principle finally shape Einstein's critical and radical attitude in that very context. Recall that during the period 1925-1931, his attitude was dominated by the following question: Is quantum mechanics a consistent theory (Heisenberg1985, Band III, 1969-1976, S. 411)? The first time he is reported to have written approvingly about quantum mechanics was when he wrote in May 1926 to Schrödinger about those advances which the latter had made in this field (A. Pais 1982, 440). For many years, Einstein believed that the theory contained logical contradictions. And he must have done so mainly in the context of Heisenberg's uncertainty relations. By 1933 he had already given up this position, openly recognizing that the theory was a logically consistent theory (A. Pais 1982, 442). By 1935, there was a turning around of the question concerning it, resulting in the Einstein-Podolsky-Rosen-Gedankenexperiment of 1935 (A. Pais 1982, 449). The question which has ever since dominated the foundational debate about the theory is the question which the Einstein-Podolsky-Rosen-Gedankenexperiment addressed in the very title of the article: "Can Quantum-Mechanical Description of Physical Reality be Considered Complete?"[39]

It might be argued that only someone who was already guided by his general conception of what makes a physical theory really a good theory - as being consistent, complete, and as possessing other virtues - could have chosen to make a new theory such as quantum mechanics the subject of a relentless critical debate. The chief advantages of the strategy followed by Einstein in his debates with Bohr, and others, seem then precisely to lie in those areas where he could simultaneously test his own general conception of a physical theory and his

---

[39]A. Einstein, B. Podolsky, and N. Rosen (1935), "Can Quantum-Mechanical Description of Physical Reality Be Considered Complete?", Physical Review 47 (777-780).

understanding of quantum mechanics, one against the other. In this sense, then, one might very well conclude, the whole debate on quantum mechanics as initiated by him, right from 1925 through 1935 to the end of his life, indeed represents a great dialogue on physical theory. In the context of a rather revolutionary scientific change, which quantum mechanics brought about, it raises the same question again and again: What kind of fundamental theory can physics aim at? And what kind of critical appraisal is appropriate in such a context?

It is quite relevant here to note how Einstein stated his points of agreement and disagreement on the interpretation of quantum mechanics in one of the last discussions Heisenberg had with him. In Heisenberg's (1985, Band III, 1969-1976, S. 411) own words:

> "I had a discussion with Einstein about this problem in 1954, a few months before his death. It was a very nice afternoon that I spent with Einstein but still when it came to the interpretation of quantum mechanics I could not convince him and he could not convince me. He always said 'well, I agree that any experiment the results of which can be calculated by means of quantum mechanics will come out as you say, but still such a scheme cannot be a final description of Nature.' "

Thus, Einstein reiterated his belief in his general conception of a physical theory, including the Heisenberg-Einstein context principle, by accepting quantum mechanics as far as it seemed to him to go while insisting on its incompleteness. The latter theme has been at the core of the Einstein-Podolsky-Rosen-Gedankenexperiment as well as the Bohr-Einstein debate.

If we seriously believe in the Heisenberg-Einstein context principle but are at the same time critical of a particular physical theory T, such as quantum mechanics, we could very well reject T's claim to being a final theory by arguing as follows: Since it is always the theory which decides what can be observed, we had better always decide first what kind of fundamental theory T should physics aim at. We had better do so before we decide which of the old concepts - wave, particle, position, velocity etc. - have a limited range of applicability in the domain of the extant theory, which limitations in the case of quantum mechanics are given by Heisenberg's relations of uncertainty. And we had better ask such an important question when it is openly a matter of finding a theory which may be taken to serve as a fundamental, even final, theory for the whole of physics.

To formulate the same argument differently, if it is always the theory which decides what can be observed, then this should have important consequences, first for our understanding of the extant physical theories - the STR, the GTR, and quantum field theories such as quantum electrodynamics, quantum chromodynamics, the Standard Model of particle physics, supersymmetry as a possible extension of the Standard Model, and so on - and for our epistemic appraisal of those theories, and, secondly, for the methodology which the physicists must follow in their search for the kind of fundamental physical theory which physics can aim at. Thus, it should have general, or rather universal, methodological consequences in the following sense. Before the physicist subjects the extant theories to severe criticism, or to epistemic appraisal, he/she had better decide first what kind of fundamental theory should physics aim at, this being the question to which every generation of physicists, and methodologists, must return, thereby making physics essentially open-ended. And given the specifications for such a theory, he/she had better decide what kind of methodology of theory-finding should it adopt or develop. But the methodology of theory-finding, like the methodology of theory-testing, can work best when it is guided by the context-building from within physics in which all the extant successful theories participate.

Many specialists in the foundations of classical physics recognize rather only indirectly the enormous power of the optical metaphor which seems to be at work when we say that it is always the theory in physics which decides what can and what cannot be observed. But I think that there is more than an optical metaphor at work here. The methodological importance of Heisenberg-Einstein context principle has rather to do with that fundamental aspect of physical theory, or its mathematical formalism, which performs best by making novel predictions and which decides whether all that which is being predicted is observable. One can here think of the quark theory which forbids observation of quarks and gluons in isolation. To quote Steven Weinberg ((1992, 184) in this context:

> "The idea that quarks and gluons can in principle never be observed in isolation has become part of the accepted wisdom of modern elementary particle physics, but it does not stop us from describing neutrons and protons and mesons as

composed of quarks. I cannot imagine anything that Ernst Mach would like less.

The theory was only one step in a continuing process of re-formulation of physical theory in terms that are more and more fundamental and at the same time farther and farther from everyday experience. How can we hope to make a theory based on observables when no aspect of our experience - perhaps not even space and time - appears at the most fundamental level of our theories? It seems to me unlikely that the positivist attitude will be of much help in the future".

While there need not always be a strictly positive correlation between that which a theory predicts for the first time and that whose observability it prohibits, the more admirable will be the theory if it itself suggests experiments to test the novel predictions which it makes. In this regard, the role of its mathematical formalism will be crucial. Moreover, if the observation of certain kinds of phenomena is forbidden by a particular theory T, the same kind of phenomena may become observable on the basis of another theory T', where T' is better than T in many other respects, besides being a successor theory in the same field. The bending of the rays of star light by the gravitational force of the sun was a phenomenon predictable from Newton's universal law of gravitation but not observable (a case of negative correlation, see Pandit 1991, pp. 265-296). It became observable on the basis of Einstein's GTR which predicted it far more precisely than did Newton's law, even suggesting experiments to test its own novel predictions. In any case, by their very nature as universal statements of prohibitions, physical theories rule out one kind of phenomena by allowing some other kind of phenomena. It is not surprising if it is always the theory which decides what can, or what cannot, be observed.

We believe that the examples of Pierre Duhem, Werner Heisenberg, Albert Einstein and Niels Bohr offer themselves as excellent candidates for case-studies in the methodology of dynamic core-contexts of development for finding the path to the kind of fundamental theory which physics at the frontier of discovery can aim at. Duhem's work on the aim and structure of physical theory - T - remains a great milestone in the methodology of physics of the last century. While reflecting on this theme, Duhem (1906/1954, p. 32) argues that there are two major aspects, or parts, of T, which call for careful attention and analysis by the physicists and methodologists of physics. On the one hand, T has an explanatory part with which it proposes to take hold of the reality underlying the phenomena. On the other hand, T has a representative part, with which it proposes to bring about a natural classification of laws (Duhem 1906/1954, 31-32). If we were to rationally reconstruct this distinction, the former aspect of T can be called its explanatory power in the standard sense of this term. The other important aspect of T, to which he draws our attention for the first time, remains still a subject of great neglect at the hands of the experts in the philosophy of physics. We believe that it deserves a serious scholarly attention. Elsewhere (Pandit 1982, 1991), we have argued for the methodology of theory-problem interactive systems, distinguishing between the explanatory power of T and the resolving power of T.[40] In this proposal, it is the resolving power of T which emerges as the most important aspect of a physical theory. A close similarity between Duhem's distinction and our own distinction cannot be ruled out. We think that in his analysis of physical theory, its aim and structure, Duhem might have been far ahead of his times, particularly in the context of the state of the art called logic and methodology of scientific discovery. In so far as his methodology clearly recognizes two parts of T, with the representative part being accorded by him the most important role in the dynamics of the growth of scientific knowledge within physics, our present methodological proposals - particularly the methodology of the dynamic core-contexts of development - can be seen as a further development of Duhemian methodology. On the other hand, we think that both the Bohr-Einstein debate and Heisenberg-Einstein dialogue on physical theory should be studied as further milestones in the methodology of physics leading to quantum core-context-building. Together with Einstein's theories of relativity, directly or indirectly they help in building up the core-contexts of development in contemporary physics, if only by promoting an improved understanding of quantum mechanics. Equally important and relevant in this con-

---

[40]In a discussion with Lorenz Krüger, going back to May-June 1986 at the Freie Universität Berlin, where one of the authors had just begun his research stay as a Fellow of the Alexander von Humboldt-Stiftung, Krüger cited Dirac's prediction of the positron, in agreement, to show how by its resolving power a physical theory is able to generate new problems. Dirac's relativistic theory of the electron predicted the existence of the positron before its experimental discovery by Anderson. In this case, it is clearly a whole cluster of theories - quantum mechanics, the special theory of relativity and the hypothesis of the electron spin - which determined the problem which Dirac was so successful in solving.

text are Heisenberg's and Einstein's conceptions of physical theory, in which there is no role for *ad hoc* hypotheses as a possible strategy to adjust the parameters of a theory which is faced with unfavourable experimental evidence. Thus, Heisenberg's (1948, 331-336) geschlossene Theorien[41] and Einstein's (Einstein, Podolsky, Rosen 1935, 777-780) complete theories may even be regarded as variations on the same theme. But it must be noted that they share nothing in common with the Kuhnian paradigms. Heisenberg's intention in introducing the concept of geschlossene Theorien is echoed by Steven Weinberg's (1992, 88) following statement about quantum mechanics:

> "I simply do not know how to change quantum mechanics by a small amount without wrecking it altogether."

Heisenberg also held the view that new problems in physics are inherited by the physicists from its historical development. They cannot be invented as theories are. We think that this asymmetry between problems and theories is a resource of great methodological significance in so far as this too hints at the role of context-building in physics. The same is, we think, true of what Einstein (Einstein 1919, 1960, 77) said as early as 1916:

> "Es ist das schönste Los einer physikalischen Theorie, wenn sie selbst zur Aufstellung einer umfassenden Theorie den Weg weist, in welcher sie als Grenzfall weiterlebt".

To turn to quantum mechanics as a core-context of theory development in current physics, and not just as a tool of description or

---

[41]We reject as untenable the view which regards the Kuhnian notion, or rather notions, of a paradigm as peculiarly reminiscent of Heisenberg's notion of a "closed theory". The pragmatic complexities and the semantic ambiguities inherent in the former, which have led Kuhn himself later on to distinguish the different senses of 'paradigm' present in his own work, do not allow any comparison between them. A Kuhnian paradigm in crisis, for example, is not just a theory in crisis. It is rather the normal scientific community and its disciplinary matrix which are in crisis when there is such a crisis at all. In other words, when a science, or any part of it, is undergoing test, it is not the theory which bears the main ordeal of the test. It is rather the puzzle-solving normal scientific community - science in context - which bears the ordeal. Thus, in Kuhn's picture, the theory suffers a displacement from that center which it still occupies in actual scientific practice and in the methodology of science. When Heisenberg argued that a closed theory is not amenable to small improvements by small changes in its structure - in its formalism - he was only following the best of the traditions in scientific practice, which disallows any scientific change by resort to ad hoc hypotheses. Many physicists would follow Heisenberg's own idea of a "closed theory" as a methodological rule facilitating rational scientific change within physics by replacing a theory as a whole whenever it is possible to do so. My criticism here applies particularly to the kind of account to be found in Mara Beller (1999), Quantum Dialogue. University of Chicago Press: Chicago & London. P. 288. See Thomas Kuhn (1962), The Structure of Scientific Revolutions. University of Chicago Press: Chicago, 2nd enlarged ed. 1970.

calculation, is a most challenging as well interesting task for physics at present. If there has been any turning around of fundamental physical theory to allow it to play such an important role, it is to be found in quantum mechanics, besides other theories. Did not Duhem (1906/1954, 32) already hint at that very aspect of physical theory which can suggest discovery? One of the authors has himself proposed elsewhere (Pandit, 1982, 1991) that serious attention ought to be paid to the resolving power of a physical theory. It was not easy to realize it then that the proposal is essentially a variation on the Duhemian methodology, or rather on the Duhem-Heisenberg-Einstein methodology. To our question which was posed above, there emerges an answer from the fore-going discussion as follows:

> The kind of fundamental theory which physics can aim at, as it moves from one frontier of discovery to the other, is that which not only explains what there is to explain in physics but also determines new problems on its way to interrogating nature as far as its own framework allows it to do.

As cited in (Chandrasekhar 1974, p. 17), according to an aphorism of Eddington "You cannot believe in astronomical observations before they are confirmed by theory." We think that what is true of astronomy is equally true of cosmology and the elementary particle physics. For in these fields the two, theory and observation, enjoy a kind of strong reciprocal relationship. A sound cosmological model of the very early universe will receive important inputs from what happens at the frontier of discovery in high energy physics. Think of a fundamental theory in physics unifying all the four fundamental forces. Such a theory should find its testing ground in the very early universe as described by a plausible cosmological model of the various epochs of the universe.[42] From this, taken together with the Heisenberg-Einstein context principle, we can make a strong case for a rather strongly reciprocal relationship between theory and experiment in the disciplinary contexts of physics, astronomy and cosmology as follows: No only is it always the case that it is the theory which decides what can be observed, or measured, but also that it is the theory

---

[42]The kind of reciprocity we are here hinting at is expressed in the following statement by Edward W. Kolb and Michael S. Turner (1990, p. 494): "Even a less than optimist person would have to conclude that the answers to many of the pressing questions must lie in understanding the earliest history of the universe, which in turn necessarily involves the application of physical theory at the most fundamental level to the cosmological setting." Turn to (Pandit 1991, pp. 265-326).

which is invited to interpret observations before those observations can be confirmed and allowed to play a rather crucial role in the frontier of scientific discovery. On the other hand, every new theory must be confirmed by novel experiments, such that the larger the number and variety of such experiments, the more mature will the theory be as dynamic core-context of development.[43]

---

[43]The celebrated astrophysicist S. Chandrasekhar argued for a similar view in the context of astronomical observations. Turn to (Chandrasekhar 1974, p. 17).

Chapter 7

Physics in Core Context (by H G D)

Cette harmonie que l'intelligence humaine croit découvrir dans la nature, existe-t-elle en dehors de cette intelligence? Non, sans doute, une réalité complètement indépendante de l'esprit qui la conçoit, la voit ou la sent, c'est une impossibilité. ... Mais ce que nous appelons la réalité objective, c'est, en dernière analyse, ce qui est commun à plusieurs êtres pensants, et pourrait être commun à tous; cette partie commune, nous le verrons, ce ne peut être que l'harmonie exprimée par des lois mathématiques. (H. Poincaré (1905), p. 9)

In the preceding chapters we have used physical examples as illustrations for the epistemological research program exhibited. This made the presentation of the physical background necessarily fragmentary. In this chapter we want to present important aspects of contemporary physics in context, namely the present Standard Model of Particle Physics and that of Cosmology. Though we refrain from going seriously into the mathematical foundation, we want nevertheless at least to give an idea of the underlying concepts. Therefore we start this chapter with a section of one of the most important concepts of contemporary physics, namely that of gauge symmetry. This principle was developed by the Mathematician, Physicist and Philosopher Herman Weyl [1]. The concept of Gauge Symmetry was constructed in connection with Einstein's Theory of General Relativity, which turned out to be a special example of a theory fulfilling this symmetry. Later it was realized that is was of eminent importance in quantum physics, especially relativistic quantum field theory, and thus to be the foundation of the Standard Model of Particle Physics. This chapter is based on Dosch(2008) and Dosch(2007).

---

[1]For the influence of philosophy on Herman Weyl see Sieroka 2010

7.1 Symmetries and the Gauge Principle

Symmetries have played an important role in physics for a long time, but they became of paramount importance in quantum mechanics and particle physics. Hermann Weyl and Eugene Wigner were the first to recognize that. One can say that symmetry principles were the principal ingredients in the development of the Standard Model of Particle Physics and also in the attempts to go beyond it. Since in this book we are largely concerned with principles of theory development it is useful to go through some rather formal considerations concerning symmetries though of course we cannot use the full highly polished formalism of mathematics related to this field.

7.1.1 Symmetries in General

Before we come to the concept of gauge symmetries, we have to discuss some properties of symmetries in general, especially their mathematical properties essential for application in physics. Herman Weyl has written a beautiful non-technical book on symmetries (Weyl(1952)).

Some Mathematical Formalism, in Words

In thinking of symmetries, one first associates a rather passive and perhaps admiring attitude, but symmetry becomes apparent only through an action, the symmetry transformation. Think of an apparent example of symmetry, a butterfly with extended wings. The observer is charmed by finding out that both wings are congruent. In order to see that, he has – at least in his mind – to turn one above the other. The symmetry operation here is the reflection at the axis of the butterfly. The manufacturer of an kaleidoscope makes use of several reflections in order to produce a symmetrical pattern.

The rotational symmetry plays an especially important role, going far beyond intuitive applications. If we rotate a wheel with four spokes by an angle of 90 degrees, it comes again to congruence, see figure 7.1 a). Its form is invariant under a rotation of an integer multiple of ninety degrees. The massive wheel of a prehistoric oxen chart, see 7.1 b) comes to congruence after a rotation by an arbitrary angle. In the case of the wheel the symmetry is called discrete, since only certain angles lead to congruence, in the case of the massive wheel the symmetry is called continuous.

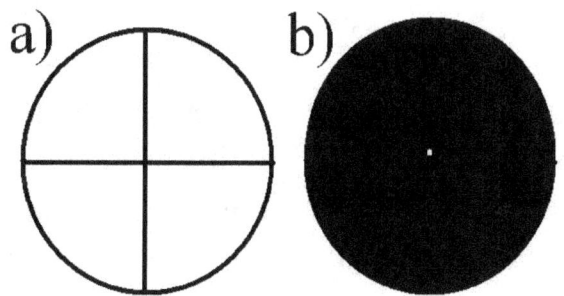

Figure 7.1: Illustration of a discrete and a continuous symmetry: a) Wheel with four spokes. After a rotation by $360/4 = 90$ degrees it becomes congruent. b) A massive wheel, it comes to congruence after a rotation by an arbitrary angle.

Rotations in space are intuitively accessible to everybody and it is nice that we can study with them essential features of many symmetry transformations. There are three essential properties:

1. Two rotations performed successively result again in another rotation. If I turn myself around my own axis and then make a handstand I have performed two successive rotations, an acrobat could have done that in a single action. We say : the product of two rotations is again a rotation.

2. We can undo a rotation by turning back. One calls the rotation which undoes the first one its inverse. Thus to each rotation exists its inverse.

3. The third property seems very trivial, but we have to mention it since it is important for abstract mathematical considerations: If one performs three rotations successively, it does not matter if one groups the first two or the last two to one rotation.

A set of mathematical objects having these three properties is called a *group*. Therefore the colloquial term 'group of rotations' has also a deeper mathematical meaning. Rotations form indeed a group with the self-explanatory name 'rotational group'.

Another important group of transformations are the translations, shifts in space and time. These are related to symmetries too, namely to those of crystals; if a crystal is shifted exactly by the distance of two atoms it becomes, up to the border regions, congruent. We assume that space and time are homogeneous, that is all space and time points are equivalent. This we can consider as a continuous symmetry, since space is congruent after an arbitrary shift in space and time. This means that the outcome of an experiment is – with all other conditions unchanged – independent of the place where, and of the time when it was performed. This invariance of physical processes under space and time translations has far reaching consequences, it leads to conservation of momentum and energy; more precisely: the invariance allows to define quantities, which we call momentum and energy and which are conserved. The relation between symmetries and conservation laws is a consequence of a mathematical theorem by Emmy Noether.

The group of translations is 'commutative', it does not matter if I go first five steps in forward direction and then three steps sidewise to the right or if I first go three steps sidewise to the right and then five steps in forward direction, in both cases I come to the same place. On calls such commutative groups after the Norwegian mathematician N.H. Abel 'Abelian groups'. The group of rotations is not commutative. Our position is different if I first turn myself by 180 degrees around my own axis and then fall onto my nose or if I first fall and then turn myself. The groups in which the result depends on the order of the operations are called noncommutative or non-Abelian. The mathematical structure of both kinds of groups is quite different and leads to far reaching physical consequences, as we shall see later.

Before we continue with mathematics, we briefly disgress into how symmetries are checked in physics. we have mentioned earlier that the outcome of an experiment does not depend on the place where it is performed, but have added 'with all other conditions unchanged'. To realize unchanged conditions is sometimes quite difficult, but slightly changed conditions can often be taken into account by calculated corrections. We all have learned that for small oscillations the period of a pendulum depends only on its length. But a very precise pendulum clock brought in 1672 from Europe to Cayenne was there a bit slow. Huygens could explain this phenomenon which at first sight contradicts the homogeneity of space. Near the equator the gravitational force of the earth is diminished by the centrifugal force due to the rotation of the earth which in turn leads to an exactly calculable increase of the oscillation period. Since Huygens had made use – at least implicitly – of the homogeneity of space, the exact agreement of his calculation with the observed difference was a corroboration of this hypothesis. A bit more subtle are the temporal fluctuations of certain experimental results at the big particle accelerator at CERN. They can be explained by the tidal forces of the moon which are strong enough to deform the accelerator in such a way that the energy of the particles changes slightly.

But let us come back to rotations, where we still can learn some mathematical concepts. Up to now we have considered rotations in a purely intuitive geometrical way. But in order to make numerical calculations we have to get also an algebraic grasp of it. This is done by the so called 'representation of a group'. We can represent each point of a body by three coordinates. If the body is rotated the coordinates change, but not arbitrarily but in a well defined way, determined by the special rotation. This change of coordinates can be represented by a quadratic scheme of nine numbers, a '3 × 3 matrix'. we are not going to explain the algebra of matrices, who knows it does not need to learn it, and who does not know it, will not learn it here. In our context it is only important that each rotation can be represented by a matrix and that this possibility of representation by a matrix holds for many groups. Very often it is possible to classify groups by their simplest matrix representations.

Therefore one can introduce rotations purely algebraically, namely as the group of 3 × 3 matrices, which leave lengths and angles unchanged. It might seem natural to represent rotations in three dimensional space by 3 × 3 matrices, but there are also representations by matrices of higher dimension, for instance 6 × 6 matrices. They rotate more complex objets as fluxes of force or stress tensors. Generally, the size of a matrix determines the number of objects in a 'multiplet' which are transformed by the transformation. A $d \times d$ matrix acts on a multiplet of d elements, one calls d the dimension of the representation. The simplest, the 'fundamental' representation of rotations in space is of dimension $d = 3$, corresponding to the three coordinates of a space point.

For each transformation there exists a particularly simple one, which is called trivial. If a quantity remains unchanged under a rotation, e.g. the volume of a body, one can also say: With respect to this quantity all rotations are represented by multiplication with the number 1, and this identical operation is the 'trivial representation' of the group. Its representation matrix is a 1 × 1 matrix which has only one entry, just the number 1. Quantities transforming under the trivial representation, that is which do not change at all, play a special role, since they show a particularly high degree of symmetry. If in the following we speak of the 'simplest' representation then we mean the simplest non-trivial one.

We have to introduce another important concept, that of the 'generators'. In order to determine a rotation uniquely we need three angles and *vice versa* by these three angles the rotation is uniquely fixed. In order to construct a rotation, we hence need three 'generators' which determine how to construct the rotation with the three angles. Grossly speaking the three generators of the rotational group are obtained from rotations by infinitesimally small angles. We denote the generators for rotations around the x-, y-, and z-axis by $\mathbf{L}_x$, $\mathbf{L}_y$ and $\mathbf{L}_z$; out of these three generators all rotations can be constructed. The generators meet rather simple relations which can be obtained from the geometrical properties of rotations. We have e.g.

$$\mathbf{L}_x \cdot \mathbf{L}_y - \mathbf{L}_y \cdot \mathbf{L}_x = i\mathbf{L}_z.$$

Here again $\mathbf{L}_x \cdot \mathbf{L}_y$ means: apply first $\mathbf{L}_y$ and then $\mathbf{L}_x$. These relations of the generators, the so called commutation relations, nearly fix the group and there happens in quantum mechanics a wonder which we shall treat in the next section.

The concept of generators is not confined to rotations, there is a whole class of groups which can be constructed with the help of generators. These are the so called Lie groups which play an important role in physics.

We have consciously used names like 'group', 'generator', 'represent' in the meaning of every day language too; one should however bear in mind that they have a very precise meaning in mathematics.

Symmetries in Quantum Physics

One of the essential differences between quantum physics and classical physics is the role of 'observables'; in quantum physics operators are assigned to observable (measurable) quantities. The result of a measurement is determined by the result of the action of the operator on a state. In quantum mechanics as well as in classical physics the observable 'angular momentum' plays a crucial role. The operators which correspond to these important observables are essentially the generators of the rotational group.

We have mentioned shortly that the commutation relations *nearly* fix the group. By investigating which other group can be constructed from the generators with the same commutation relations one finds besides the group of rotations another one, namely one, which can be considered as rotations in a two-dimensional space. This two-dimensional space is, however, not an ordinary plane but one in which the coordinates of a point are complex numbers. This group is called $SU(2)$, SU stands for 'special unitary' which designates a rotation of points with complex coordinates, (2) stands for two dimensions. These new operators can be called *spin* operators, up to now they are only formally defined. Since the space is two dimensional, the simplest non-trivial 'spin' has two possibilities of orientation. In quantum physics, it can be shown that the possible values for the angular momentum are integer multiples of the (reduced) Planck constant $\hbar$, but for the spin the possible values are multiples of $\hbar/2$. The simplest objects transforming under this spin group $SU(2)$ have two components, hence they are doublets; they are called 'spinors'. In the following we shall quote the spin always in natural units, spin $\frac{1}{2}$ means spin $\frac{1}{2}\hbar$ and we say spin is half-integer.

All this is mathematically rigorous, on the other hand it sounds a bit like playing around with abstract concepts. But now the miracle comes: To the purely formally introduced spin operators there correspond indeed physical observables. There exist particles with half-integer spin values, and these particles are by no means exotic, but constitute the matter surrounding us: electrons, protons and neutrons. Whoever underwent a magnetic resonance tomography has profited from the spin of the proton.

If a particle has spin $\frac{1}{2}$, its spin can be oriented either parallel or antiparallel to a fixed direction and has with reference to this direction the values $+\frac{1}{2}$ and $-\frac{1}{2}$, normally represented as up, $\uparrow$, and down, $\downarrow$. If such a state is rotated by 180° around an axis perpendicular to the fixed direction mentioned above, the state with spin orientation $+\frac{1}{2}$ ($\uparrow$) changes into spin orientation $-\frac{1}{2}$ ($\downarrow$). This has consequences: If the interaction does not change under rotations, that is if it is invariant under the rotational group, then the experimental results are invariant too. This means: if a particle with spin $\frac{1}{2}$ is involved it does not matter if its spin orientation is $+\frac{1}{2}$ or $-\frac{1}{2}$ ($\uparrow$ or $\downarrow$).

Right and Left handed Spinors, Chiral Symmetry Let us generalize rotations in three-dimensional space to the space time transformations which are in accordance with the principles of special relativity! Under these transformations not only angles and lengths remain invariant, but also the velocity of light in the vacuum. This group is called *Lorentz group*, after the Dutch physicist H.A. Lorentz. Its fundamental representation is four dimensional, because of the three space and the one time dimension. The four dimensional matrices are called Lorentz transformations. With this group the miracle of spin repeats itself. There is a more general group which can be represented by 2×2 matrices, from which all representations of the Lorentz group can be constructed. This had been recognized in 1929 by the mathematicians B. van der Waerden and Hermann Weyl. Especially Weyl pointed out that for the doublets transforming under that two dimensional representation one can construct a relativistically invariant wave equation which is simpler than the Dirac equation. These doublets are now called *Weyl spinors*.

There exist two completely independent kinds of Weyl spinors – mathematically speaking there are two 'unitary inequivalent representations'. Both describe spin-$\frac{1}{2}$ particles. For one kind the spin is parallel to the direction of movement, these are the *right-handed spinors*. For the other kind it is antiparallel, these are the *left-handed spinors*, see figure 7.2. One calls this property *chirality*. Particles and antiparticles have opposite chirality: If a particle is described by a right-handed spinor, then its antiparticle is described by a left-handed one.

A particle with definite chirality must be massless, as can be seen easily. If a particle has a mass different from zero its velocity is always smaller than that of light, as a consequence of that it can be overtaken by an observer. If the observer is faster than the observed particle it moves for him backward, exactly like an overtaken car. Seen from the passing lane, a right handed particle becomes a left-handed one, since the direction of movement has changed, but not the spin. A massive particle can therefore not be described by a Weyl spinor. Because of that Dirac had to introduce spinors with *four* components. The four-component Dirac spinors are composed of two Weyl spinors, a right- and a left-handed one.

Wolfgang Pauli had written in 1933 a famous review on 'The General Principles of Quantum Mechanics', known to experts as the 'New Testament'. There he referred also to the Weyl spinors and their wave equation: 'These wave equations are however not invariant under reflections (exchange of left and right) and therefore not applicable to

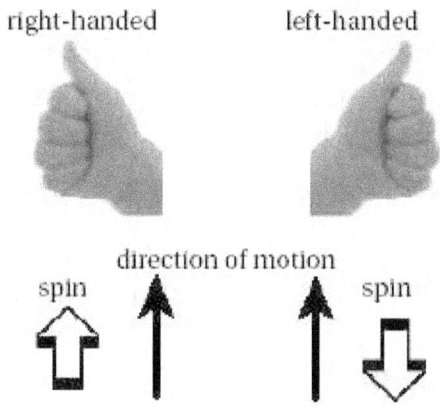

Figure 7.2: Chirality (handiness) of particles. A particle the spin of which is parallel to the direction of movement is called right-handed. If spin and direction of movement point in opposite direction, is is called left-handed. The thumb indicates the direction of movement, the other fingers the rotational movement of the spinning particle.

physical reality. The absence of invariance of the wave equation manifests itself in a peculiar coupling of the direction of the spin-angular momentum and the current'.

The argument seemed to be valid at the time, but after one had realized that weak interactions are not invariant under space reflections the curse of Pauli changed into a blessing. Nowadays the description in terms of Weyl spinors is considered to be more fundamental than that in terms of Dirac spinors. The latter can indeed be constructed from Weyl spinors.

One calls transformations which transform right- and left-handed particles separately *chiral* transformations and the corresponding symmetry *chiral symmetry*. Chiral symmetry is therefore a symmetry which requires spin $\frac{1}{2}$ particles to be massless. This plays an important role in the discussion of mass generation in the Standard Model of Particle Physics and its supposed extensions.

Internal Symmetries

It is quite remarkable that physical observables correspond to these mathematical concepts, which have found even ample technical applications, e.g. in medical diagnostics. But there is also a second miracle around the door: Nature has made extensive use of the group $SU(2)$ a second time, as a so-called internal symmetry. Internal symmetries refer to properties which have no connection to behavior under space-time transformations. The discovery of internal symmetries is one of the most remarkable feats of particle physics. It started with the detection of the neutron by Chadwick in 1932. Immediately after that Heisenberg guessed that the atomic nucleus is composed of protons and neutrons and not – as assumed previously – of protons and electrons. The masses of proton and neutron are quite similar, the difference is only about one per mil, and from properties of the nucleus one could conclude that the binding forces of neutrons and protons must be quite similar too. Therefore Heisenberg proposed already in 1932 that the proton and neutron are related by a symmetry transformation. In this respect proton and neutron are the different states of one and the same particle which differ from each other only by their orientation in an 'internal symmetry space'. The analogy with the spin was obvious: The electron can appear in two spin states too, and nobody would speak of two different particles, but of two different states of one and the same particle. Since 1941 the proton and neutron are summarized under the name 'nucleon'.

Heisenberg had introduced the name ρ-spin for the distinctive properties of proton and neutron, today the name 'isotopic spin' or shortly 'isospin' is in common use. One calls the internal symmetry space 'iso space' and says that the proton has the isospin orientation $+\frac{1}{2}$, the neutron $-\frac{1}{2}$. Already Heisenberg had used the two-dimensional matrices introduced by Pauli to describe spin; they are the generators of the group $SU(2)$.

The formalism of Heisenberg was regarded as clumsy and of little help, however, and was for a few years nearly forgotten. It was revived only in 1936 by B. Cassen and E. Condon, who were able to interpret with its help new experiments on charge independence of nuclear forces. They fully made use of the formal identity of spin and isospin. As in spherically symmetrical situations results are unchanged under rotations, so nuclear forces were assumed to be independent of rotations in iso space. These 'rotations' transform a neutron into a proton as rotations in ordinary space transform an electron with spin orientation $-\frac{1}{2}$ in one with orientation $+\frac{1}{2}$. We want to emphasize however that the rotation in iso space is a purely formal operation and has no direct intuitive meaning.

Nevertheless we can transfer the *formal* results of the the behavior of two spins directly to the behavior of two nucleons. For that purpose we have only to replace the spin orientation $+\frac{1}{2}$ ($\uparrow$) by a proton, and the orientation $-\frac{1}{2}$ ($\downarrow$) by a neutron.

These considerations got Nicholas Kemmer to go a decisive step further. He attributed this symmetry, which was introduced originally only to classify states, to the dynamics of meson exchange. The meson theory of nuclear forces, proposed by Yukawa in 1934 (see sect. 3.3) got in 1937 an enormous impetus. S.H. Neddermaier and C. D. Andersen had discovered a new kind of particles with a mass of ca 240 electron masses and the electric charges $+1$ and -1. This particle with a mass between the electron and the nucleon mass was first called 'mesotron'. In the year 1937 the paper of Yukawa was quoted for the first time in a western journal. R. Oppenheimer und R. Serber proposed that the new 'mesotrons' detected by Anderson and Neddermeyer are the particles postulated by Yukawa. Only positively charged and negatively charged mesotrons had been observed, but in order to obtain symmetry, Kemmer had to make a daring hypothesis: he had to postulate the existence of a third, a neutral, mesotron, which together with the charged ones forms a triplet with isospin 1 (iso triplet). With these assumptions he could construct an interaction between the isospin doublet of nucleons and the isospin triplet of mesons which was invariant under rotations in iso-space. He could prove that this invariance pertains also to the interaction of nucleons among themselves.

So the first 'internal symmetry' was introduced in particle physics, that is a symmetry which is independent of the space-time behavior of particles. It was modelled after the example of spatial rotations and the (quantum mechanical) formalism developed there could be directly applied to the isospin.

Internal symmetries play nowadays a decisive role in particle physics and we shall come back to this subject in the section on the Standard model. The invariance under rotations in iso-space could not be extended to all interactions, since the electromagnetic interactions of the proton and neutron are very different. Therefore this invariance was only assumed to be valid for strong interactions, the interactions responsible for the forces inside the atomic nucleus. One says that the isospin symmetry is 'broken' by the electromagnetic interaction. This breaking of the symmetry was also used as an explanation for the fact that the proton and the neutron have not exactly the same mass. The small mass difference should be due to electromagnetic interactions, though it seemed somewhat mysterious that the neutron is heavier than the proton. From classical arguments one would have concluded that the electrostatic repulsion inside the proton makes it heavier than the neutron. The breaking of the symmetry could also explain why one could first not detect the neutral mesotron postulated by Kemmer: it could decay by the electromagnetic interaction into two photons. S. Sakata und Y. Tanikawa had calculated in 1940 the mean life time of the neutral particle and found a time of about one hundred millionth nanosecond (10^{-17} seconds). This lifetime is about hundred billion times smaller than that of the charged mesotrons and therefore it could not be detected by the methods of Neddermaier und Andersen, it was found only in 1949.

The group $SU(2)$ plays also an important role in the Standard model of particle physics, see next section, namely in the sector of weak interactions. Another important group in the Standard model is $SU(3)$. It is based on 3×3 matrices and has 8 generators.

Discrete Symmetries

Discrete symmetries are not particularly discreet, but the related transformations do not depend on a continuously varying parameter, as we have mentioned in the beginning of the section. The discrete symmetries we shall discuss now are:

- space reflection or parity transformation **P**. In this transformation each point of space is reflected at the origin, that is the coordinates of a point change their signs.

- Time reversal **T**. Here the direction of time is reversed. It is like running a movie backwards.

- Charge conjugation **C**. Here particles are transformed into antiparticles, an electron, e.g. is transformed into a positron. This transformation plays a role only in quantum field theory.

Though we have little experience with space reflections at a point, we all are familiar with the reflections at a plane from the mirror. You can convince yourself that your right hand in a mirror looks like your left hand before the mirror. This is also true for space reflections at a point, a right hand is transformed into a left hand.

From everyday-life we have no experience of time-reversal symmetry, growing older is a normal process, getting younger happens only in science fiction. Nevertheless the laws of mechanics and classical field theory are invariant under space reflection and time reversal. This was also first assumed to be true in particle physics but our eyes got opened in the course of time. we shall come back to this subject later in detail.

One can assign to a field an internal parity. If the components of the field change their sign under space reflection, the parity is called 'odd' or -1, if they stay unchanged, the parity is called even or $+1$. This parity is also assigned to the particles which are the quanta of the field. Under space reflections the components of the electromagnetic field change sign, the photons have thus the parity -1.

There exist also oriented quantities which do not change their direction (that is the sign of the components) under space reflections, they are called pseudo- or axial vectors. Spin and angular momentum are examples of axial vectors. For particles with half integer spin the parity of the antiparticle is opposite to that of the particle. The total parity of a state is the product of the internal parities and the 'external parity', the latter is determined by the angular momentum. If an interaction is invariant under space reflections the total parity remains unchanged for all reactions due to this interaction.

In the same way as parity with respect to space reflections one can introduce an internal charge parity (or C-parity) with respect to charge conjugation $\mathbf{C}$. Even if a particle is its own antiparticle the sign of its quantum field can change under charge conjugation. This is the case for photons, we therefore assign the photon the charge parity -1. Only particles or states which are neutral can have a definite charge parity, for charged particles one can introduce the so called G-parity, which is a combination of charge conjugation and a rotation in iso-space. However, we shall not discuss this here. If invariance under charge conjugation holds, the dynamical laws are the same for the world and the antiworld and the product of charge parities is conserved. That means, even if in some reactions particles and antiparticles are annihilated, and completely new particles are created the product of the internal charge parities of the final state is the same as that of the initial state.

There are two further internal quantum numbers, the 'baryon number' and the 'lepton number', the introduction of which was forced by experimentally observed conservation laws. None of the classical symmetries like conservation of energy, momentum, charge and so on forbids the decay of a proton into a positron and photons. If those decays were possible with a mean lifetime not much longer than that of the universe, a large part of matter would have been decayed. Since up to now these decays have not been observed one introduces a new quantum number, namely the baryon number, B. Mesons, photons, electrons have baryon number 0, nucleons (proton and neutron) have $B = 1$. Lepton number L is formed analogously, it is $+1$ for electrons and neutrinos and -1 for their antiparticles. If a neutron decays into a proton, electron, and anti-neutrino both baryon and lepton number are conserved: in the initial and final state we have $B = 1$ and $L = 0$. In the above mentioned but not observed decay of a proton into a positron and photons, baryon number would change from $B = 1$ to 0 and hence it would not be conserved.

7.1.2 Gauge Symmetries

All fundamental interactions known today exhibit a special kind of symmetry, which for historical reasons bears the name "Gauge Symmetry". These symmetries are especially important since they do not only allow a classification of states, as all symmetries do, but they also fix the dynamical behavior of the theory invariant under that special symmetry.

Gauge symmetries have two roots: One which is rather pragmatic and stems from electrodynamics and quantum mechanics, and a deeper one which comes from general relativity and is forever connected with the name of Hermann Weyl.

In the electrodynamics of the 19th century the electric and magnetic potentials played a major role. In the attempts to explain electrodynamical phenomena by mechanical models of the ether, one was trying to deduce these potentials from ether properties. Besides the electric potential, familiar to all of us because the difference of the electric potential is the electric tension, one had also introduced a magnetic potential, a so called vector potential. From the potentials the electric and magnetic fields can be calculated easily. The fields are directly measurable via the forces they exert. It would be to long-winded to go into the history of electrodynamics; we shall only give a short description of the situation at the beginning of the 20th century. It turned out that the decisive quantities were not the potentials but the electric fields. These appear in the fundamental equations of electrodynamics, the MaxwellMaxwell, J. C. equations. The potentials seemed to be only very convenient auxiliary quantities for solving these equations. The electric potential and the three components of the magnetic potential can be united to a four-vector which has simple properties under the transformations of special relativity. In the following we shall call this four-vector shortly 'the electromagnetic potential'. The electromagnetic potential is not uniquely determined, different potentials can lead to the same electromagnetic fields and hence to the same phenomena. A change of the potential which does not change the electromagnetic field is nowadays called a gauge transformation or 'a change of gauge'. The resulting symmetry seems rather banal: A gauge transformation of the potentials transforms only the quantities, which are not directly observable, but leaves the fields, the observable quantities, unchanged.

The potentials played nevertheless an important role in the formal treatment of classical mechanics. If one wanted to incorporate the electromagnetic forces into the refined formalism of higher mechanics, one had to insert the potentials into the fundamental expressions and not the fields. This was important in the transition from classical to quantum mechanics as we shall see soon.

Before coming to quantum mechanics we want to mention shortly the second root of gauge symmetry. The starting point of general relativity is the equivalence of all coordinate systems. There is nothing like a 'fundamental coordinate system', privileged over all other systems. Therefore the directions of two vectors (that is directed quantities) at different points $\vec{P}_1$ and $\vec{P}_2$ cannot be compared directly. For that purpose we have to 'parallel-transport' the vector from point $\vec{P}_1$ to point $\vec{P}_2$, or the vector from $\vec{P}_2$ to $\vec{P}_1$. In Euclidean geometry, which is familiar to us and determines our intuition, this is very simple, but in general the parallel transport of the vectors depends on choice of the path. If the vector is transported on two different pathes to the same point, the directions might have changed, as shown in Figure 7.3a.

The deviation of the two directions is a measure for the so called curvature of the space and in general relativity it is determined by gravity. The length of a vector, however, remains unchanged, that is there exist rigid gauges. This is the so-called Riemannian Geometry.

Weyl went one step further in these considerations: He assumed

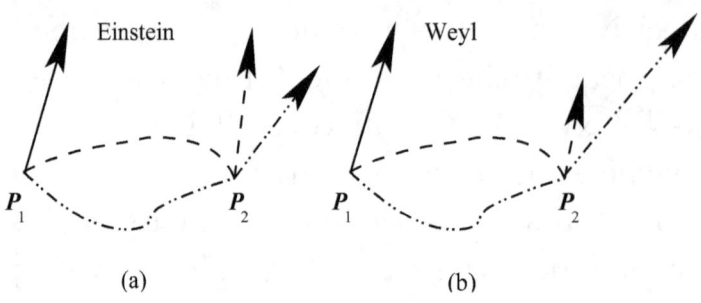

Figure 7.3: Parallel transport of a directed quantity (**a**) according to Einstein's general relativity (**b**) according to Weyl's 'near-geometry'. Einstein has as starting point Riemannian geometry where the direction may change along the path, but the length remains unchanged. In Weyl's near-geometry also the length can depend on the path

that also the length can change during a transport, see figure 7.3b. He saw this approach, which he called 'near-geometry', as a natural extension of Riemannian geometry and therefore also of the theory general relativity. Weyl did not mean that total arbitrariness prevailed in gauging length. The path-dependent modification of the length must have a physical reason. In general relativity the path-dependent change of the direction is determined by the gravitational field. One needs to know it, in order to parallel-transport a vector. Conversely the mathematical formalism of parallel transport incorporates a field which can be interpreted as the gravitational field. This is the famous 'geometrical interpretation' of gravity in general relativity.

Weyl could show that in case of a path dependent length a further field had to be introduced. This field, necessary in his 'near-geometry' had just the properties of the electromagnetic field. Therefore Weyl believed to have found with this consequent extension of general relativity an unified geometrical theory of gravity *and* electrodynamics.

The theory of Weyl was generally considered as ingenious, but it met scepticism as far as the physical relevance was concerned. Weyl was slowly convinced by his critics, among them Einstein and in retrospect he wrote: 'From the year 1918 dates my first attempt to develop an unified theory of gravitation and electromagnetism, based on the principle of gauge invariance, which we had put besides that of coordinate invariance. we have given up this theory long since after its correct core, the gauge invariance, was saved into quantum theory as a principle that relates not gravitation, but the wave field of the electron, to the electromagnetic field.' From the original application to changes in the length, however, the names 'change of the gauge', 'gauge transformation', and 'gauge symmetry' did remain.

Weyl mentions in this remark the keyword 'quantum theory'. Erwin Schrödinger was stimulated by 'Weyls world geometry' to write a paper 'On a remarkable property of quantum orbits of a single electron'. This work was still based on old atomic model of Bohr, before the invention of the Schrödinger equation. Schrödinger could in his paper not draw a clear conclusion, but it was stimulating and its significance became clear only after establishing the Schrödinger equation.

As mentioned above, the potential and not the fields appear in the fundamental expression of higher mechanics. This looks first like a violation of gauge invariance since we know that different potentials can lead to the same fields and hence to the same observable results. In classical mechanics the riddle is solved by the fact that in the equations of motion, which can be derived from the fundamental expressions only the fields appear, in a seemingly miraculous way.

In quantum theory the situation is not so simple. Here the electromagnetic potential appears directly in the Schrödinger equation and we are confronted with the problem: What happens if we change the gauge, that is if we change the potential in such a way that the fields remain unchanged? This question was answered in the same year 1926 in which Schrödinger published his equation. V. Fock from Leningrad (St. Petersburg) wrote a paper 'On the invariant form of the wave equation and equation of motion for a charged mass point'. In this paper he showed that the wave equation does not change under gauge transformations if the wave function of the charged particle is modified together with the electromagnetic potential. Observable results are not influenced by these transformations. Fock quoted the early paper of Schrödinger but he did not discuss the theory of Weyl. The great significance of Weyl's theory was however recognized immediately by F. London, who came independently to the same conclusions as Fock. The title of London's paper, published in 1927, is 'Quantum mechanical interpretation of the theory of Weyl'.

We shall not discuss this important work but come directly to the decisive publication of Weyl from 1929. For that we have to go back to the core of gauge invariance. We shall discuss it in terms of the example of the introduction of a new currency, the Euro, in several European countries. A global change of gauge is a simple matter, nothing changes, only the names. In a single country the change of the old to the new currency is an example for it. In Germany one had only to divide the prices in D-mark by 1.9558 to obtain the price in the new currency Euro. The buying power of the income was not influenced by this change (or should not have been, at least). What for one country alone looked like a global symmetry transformation turned out to be a local transformation if seen as a whole. The exchange factor 1.9558 is of little use for an Italian, if he wants to covert Lira to Euro, he needs another conversion factor, valid for Italy. For countries outside the Euro zone the conversion factor not only depends on the space (country), but also on time, it follows the course of the varying exchange rates. If we want to compare prices before and after the introduction of the new currency we need a conversion table. Only if we have checked with this table that the real prices have not changed, we can speak of a 'change of gauge'; otherwise there is a real change of the price (a physical change, so to speak).

In the old theory of Weyl the change of length was such a 'gauge transformation', in his new theory it was the change of a certain property of the quantum mechanical wave function of the charged particle. In both cases the conversion table could be calculated from the electromagnetic field. One can however reverse the roles, and this is exactly what Weyl did: If we demand gauge invariance, there must exist a conversion table and hence an electromagnetic field.

As mentioned above, the postulate of gauge invariance of length was mathematically consistent, but apparently physically not realized; but it was realized in quantum mechanics and turned out to be physically relevant there. From the principles of quantum physics alone follows a global symmetry: Since the probability is given by the square of the modulus of the probability amplitude, any change of the wave function which does not affect the modulus does not change the observable quantities. In a concise mathematical way this can be formulated in the following way: If the wave function of a state is multiplied by a complex number with modulus one, the modulus of the wave function is not changed and the results of an observation made at this state do not change. Such a factor of modulus one is called a *phase factor*.

Wave functions are extended objects and there is in principle no limitation for their extension. Weyl judged it to be unreasonable that the above introduced phase factor has to have the same value everywhere and he rejected what he called 'parallelism at a distance'. He thus postulated that the phase factor can vary arbitrarily in space

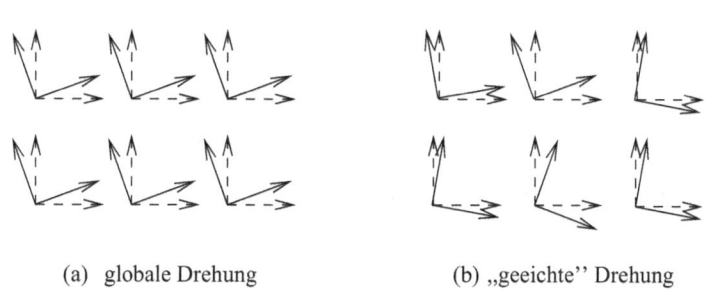

(a) globale Drehung   (b) „geeichte" Drehung

Figure 7.4: Global and local (gauged) transformation, here a rotation in the plane. (**a**) In a global rotation the rotation angle is the same at all points (**b**) In a local (gauged) rotation the rotation angle can be different at different points

and time. If this is the case a gauge symmetry holds and one needs a conversion table. The existence of a conversion table implies the existence of an electromagnetic field. The postulate of gauge invariance in quantum physics thus implies electromagnetic interactions.

Let us summarize and introduce a few technical terms useful for the following. A transformation which is independent of space and time is called global. One speaks of gauging a global transformation if one allows that the transformation acts differently at different space time points. One calls such transformations (local) gauge transformations. In figure 7.4 a global and a gauged rotation is displayed. A theory is called gauge invariant if the fundamental equations are invariant under a gauged transformation. As we have seen, the requirement of gauge invariance has dynamical consequences. In order to calculate a 'conversion table' one needs fields, the so called gauge fields; they compensate the change of the original fields in the fundamental equations.

Practically the procedure is as follows: If one begins with a theory of free, that is non-interacting, electrons the basic equation, in this case the Dirac equation, is invariant if the wave function of the electron is multiplied by a constant phase factor. One can show that this global symmetry entails charge conservation of the electron. The equation is no longer invariant if the phase factor is gauged, that if it depends on space and time. In order to restore invariance of the equation one has to modify the equation through gauge potentials. This has to be done in such a way that under a *combined* transformation of the original and the gauge potentials the equation keeps the original form. The gauge potentials are the electromagnetic potentials, not only their existence is required, but also the form of the interaction is fixed by the requirement of gauge invariance.

The multiplication with a phase factor, that is with a complex number of modulus one, is a particularly simple transformation group. A number can be considered to be a one dimensional' matrix, namely a 1×1 matrix; therefore the group of transformations consisting of multiplication with a phase factor is called $U(1)$. The 1 stands for dimension 1 and U indicates that this number has modulus one. The related symmetry is thus the $U(1)$ symmetry. We therefore say: The electromagnetic field is the gauge field of the $U(1)$ symmetry of charged particles.

In the second half of the 20th century the requirement of gauge invariance turned out to be the decisive principle for constructing new and successful theories. The adequate mathematical framework for the gauge theories is the theory of fiber bundles. The field is the whole fiber, the different potentials yielding the same field are the points on the fiber.

It isn't too far fetched to gauge also symmetries, which are more complex than $U(1)$, e.g. the groups $SU(2)$ or $SU(3)$. Attempts in this direction were made by O. Klein (1938) and W. Pauli (1953). Klein was stimulated by the work of Yukawa and came to a theory which today would be called an (explicitly) broken $SU(2)$-gauge theory. Also Pauli, very often critical towards the attempts of Weyl, was for a short time interested in this problem. He wondered whether it was possible to gauge the $SU(2)$ symmetry (isospin) in the same way as the $U(1)$-symmetry. He succeeded to construct the analogue of the electric and magnetic field but he did not pursue his investigations further; presumably he saw the difficulties to incorporate quantum physics. The problem was solved on the purely classical level, that is without addressing the problem of quantization, by C.N. Yang and R.L. Mills in 1954 and independently in a PHD thesis by R. Shaw in 1955.

In gauging the $U(1)$ symmetry one has to introduce a gauge potential, from which the gauge fields, in this case the electric and magnetic field, can be derived. The groups $SU(2)$ and $SU(3)$ are more complicated and more generators are needed to construct them, namely three and eight, respectively. The group $SU(2)$ needs therefore three and $SU(3)$ eight gauge potentials. A common feature of the gauge fields of these symmetry groups is that they can be derived from gauge potentials and that there exists an analogue of the electric and magnetic field. Therefore the field quanta of the gauge fields have spin 1 and are massless; they are called *gauge bosons*. The interaction of the gauge bosons is completely determined by the gauge symmetry and the only free parameter is the strength of the interaction. There are as many gauge potentials as the gauge group has generators. Since the group $U(1)$ has only one generator there is only one gauge potential, the electromagnetic potential.

The gauge symmetry of $U(1)$, electrodynamics, is especially simple since the product of two phase factors is independent of the order of the factors. The result of the gauge transformation is therefore independent of the order in which the transformations are performed. It does not matter if we first multiply by p and then by q or first by q and then by p. Groups with these properties are called Abelian groups. For the groups $SU(2)$ and $SU(3)$ the result depends in general on the order. The same is generally true for $SU(n)$ where n is an arbitrary integer number greater than one. These groups are called non-Abelian. One says therefore that electrodynamics is an Abelian gauge theory. Theories which are gauge invariant under non-Abelian groups like $SU(2)$, $SU(3)$ are called non-Abelian gauge theories. One calls the non-Abelian gauge theories also Yang-Mills theories.

The field equations of electrodynamics, the Maxwell equations, are linear, they contain no terms which are of quadratic or higher order in the fields. This is not only very convenient for the mathematical treatment, but has also important physical consequences: The photons, which are the field quanta of the gauge field (gauge bosons), do not interact directly with each other; they are electrically neutral. In the language of Feynman graphs this means that there is no direct coupling of three or more photons.

This is different for non-Abelian gauge theories. Here the field equations are no longer linear, there occur terms which are quadratic and cubic in the fields. This entails that the gauge field quanta couple directly to each other, one can also say that in a non-Abelian gauge theory the gauge bosons carry a 'charge' (which is of course different from the electromagnetic charge).

The coupling of the gauge bosons among each other makes the incorporation of quantum physics difficult. The path to a consistent *quantized* non-Abelian gauge theory was 'long and painful', as M. Veltman said. Of course there was no hope that one could do better than in quantum electrodynamics, therefore one tried only to establish a perturbation theory. In transposing the rules of quantum electrodynaics to non-Abelian gauge theories, that is by adding the interaction of the gauge bosons among each other to the interaction with the fermions, one experiences a bad surprise: the scattering amplitudes obtained with these graphs violate unitarity, that is probability is not conserved. This inconsistency had to be fixed up. It was achieved by new fields which do not occur at all in the classical theory. One calls such fields which only appear virtually, that is in inner lines of graphs, 'ghost fields'. Only if these ghost fields are taken into account, the non-Abelian gauge theories can be quantized in a way which respects special relativity. The technical difficulties are enormous and cannot be conveyed, therefore we only indicate the results of the final theory, which was established around 1971. Non-Abelian gauge theories are renormalizable, that is one can calculate with a limited number of parameters in principle any order of perturbation theory and therefore make very precise predictions. Many physicists contributed to the quantization of non-Abelian gauge theories, it was brought to a close by t'Hooft and Veltman.

Gauge invariance plays an essential role for the consistency of quantum field theory. Without gauge invariance one can not rule out the occurrence of more and more new terms with undetermined couplings in higher orders of perturbation theory. In particular gauge invariance seems to forbid the direct appearance of a mass term for gauge bosons. As we shall see this has stringent consequences for the Standard model. C. N. Yang reports that in a seminar in Princeton in spring 1954 he was pressed so hard by Pauli with questions concerning the mass of the gauge boson that he interrupted his talk. He could continue only after the intervention of the discussion leader Oppenheimer.

Non-Abelian gauge theories met a lot of interest in the early 1960s, that is before the questions of quantization and renormalization were settled. After the discovery of strongly interacting particles of spin 1 (vector mesons) it seemed likely that these are the gauge bosons of strong interactions. Thereby it was important to solve the problem with the mass, since all these vector bosons had a mass of more than $700 \text{ MeV}/c^2$. On the other hand the vanishing mass of the gauge bosons seemed to be an essential feature of the theory. The symmetry had therefore to be broken in some way. In the theory of strong interactions these considerations brought no progress, but they led to a break-through in weak interactions which started in 1967 with a paper of S. Weinberg.

To understand this we have to learn another important concept yet, that of the so called spontaneous symmetry breaking.

7.1.3 Spontaneous Symmetry Breaking

The limit between a broken symmetry and no symmetry at all is fluid, like that between a badly-preserved castle and a well-preserved ruin. Therefore a broken symmetry is always somewhat suspect. There is however a very respectable and well defined way of symmetry breaking, the so called *spontaneous* symmetry breaking; this will be treated in this section.

In quantum physics of fields there is a privileged state, the ground state or *vacuum* state. From it all other states can be constructed by application of field operators. This is even rooted in the axiomatic formulation of quantum field theory, the Wightman axioms. A symmetry is called *spontaneously* broken if the fundamental equations are invariant under the symmetry, but not the ground state, that is the state of minimal energy.

Spontaneous symmetry breaking is most easily illustrated in solid state physics. Here the concept was first introduced in 1928 by Heisenberg. Let us consider a magnet consisting of atoms with spin (Heisenberg magnet). Spin is a directed quantity; the interaction of two neighboring atoms depends only on the angle between the two spin directions. The interaction energy is minimal, if the two spins are parallel. This spin-spin interaction is invariant under (global) rotations since such a global rotation does not change the angles, since all spins are rotated by the same angle. The total energy is minimal if all spins are parallel. Therefore in the ground state *all* spins point in the same direction. It does not matter however in which common direction they point; therefore there are infinitely many ground states, there is one for each direction. In figure 7.5 two possible ground states are displayed. In such a case one speaks of a 'degenerate ground state' or a 'degenerate vacuum'. For a really existing magnet the spins have to point in some direction, this direction is then privileged and therefore rotational symmetry is 'spontaneously broken'.

The degeneracy of the ground state has a far reaching consequence: Since no energy is needed to make a transition from one ground state to another one, there exist massless 'particles' (in solid state physics one calls them quasi-particles), which induce the transition between

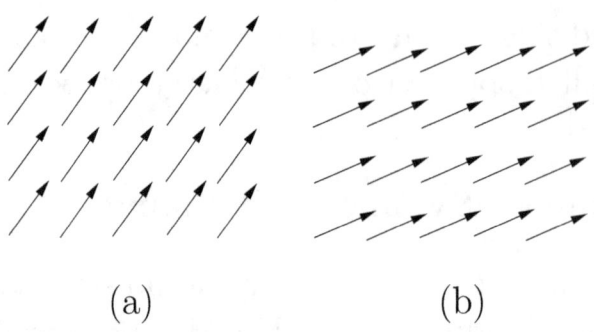

(a) (b)

Figure 7.5: A system with infinitely many ground states, a magnet of iron (ferromagnet). The energy is minimal if all spins, indicated by arrows, point in the same direction; in which direction the spins point, (**a**) or (**b**), is irrelevant for the energy

two different ground states. This is a very important general consequence of spontaneous symmetry breaking. If a continuous symmetry is spontaneously broken then there exist massless particles. This theorem was established by J. Goldstone in 1961; the massless particles, which must have integer spin, are called 'Goldstone bosons'. The Goldstone bosons of the ferromagnet are the so called magnons, the field quanta of spin waves. Better known are the field quanta of sound waves, the phonons, which can be interpreted as Goldstone bosons of spontaneous symmetry breaking of translational invariance.

In elementary particle physics the ground state is not such an intuitive concept as in solid state physics, but the mathematical treatment is the same. In particle physics the ground state is generally called the 'vacuum state' or 'vacuum'. It plays an important role in quantum field theory. The first proposal to apply the concept of spontaneous symmetry breaking also in particle physics goes also back to Heisenberg (1959). The non-linear spinor theory of Heisenberg, to which it was applied, was however not very popular, and only the papers of Y. Nambu, which appeared about two years later, were noticed.

We shall investigate spontaneous symmetry breaking by means of a simple but very important field theoretical example. For that we consider (provisionally) a particle without spin and without charge. Such a particle is described by a real field F; the field F depends on space and time coordinates, but this is not essential for the moment. The field shall interact with itself and the field quanta have the mass m. The mass m contributes to the field energy by the quadratic term $m^2 \cdot F^2$, it corresponds to the rest energy. The self-interaction yields a contribution with a fourth power of the field, $g \cdot F^4$, the positive coupling g describes the strength of the interaction. The total static

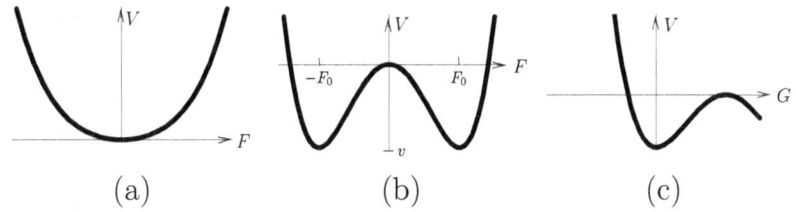

(a)   (b)   (c)

Figure 7.6: Static field energy. (**a**) With a positive quadratic term (mass term); (**b**) with a negative quadratic term leading to a sponataneous symmetry breaking; in (**c**) the field is displaced by the value F_0, $G = F + F_0$.

field energy is thus given as:

$$V(F) = M \cdot F^2 + g \cdot F^4,$$

where $M = m^2$.

The static field energy is displayed in figure 7.6a, as a function of the field strength F; one sees that it is invariant under a change of sign of the field strength F. The ground state, the state of minimal energy, is at $F = 0$. The situation is different, if we ignore that M is the square of a mass and chose it to be negative. This sounds a bit strange – and indeed it is – but let us see first what we can learn from the formalism. If it is reasonable, we should be able to interpret the result reasonably.

At small values of F the now negative term $M \cdot F^2$ is dominant and the field energy is negative; at larger values of F, however, the interaction term $g \cdot F^4$ wins and we obtain the curve displayed in figure 7.6b. Through the interplay between the 'repulsing' quadratic term and the 'attractive' interaction term the state of minimal energy comes out to be not at $F = 0$ but at a finite value $+F_0$ and $-F_0$, with $F_0 = \sqrt{-M/(2 \cdot g)}$ (please note that $-M$ is here positive). Here there exist two ground states (vacua) which do not have field strength 0, but $+F_0$ and $-F_0$. The field energy V has in both vacua the same value. The minimal value of the field energy is $v = -M^2/(4 \cdot g)$. In order to quantize the system one has to start from a definite ground state (vacuum). Therefore the symmetry under the sign change of F is spontaneously broken. It is valid for the fundamental equations (here expressed through the potential field energy), but not for the ground state choosen.

As a next step we introduce a new field, namely $G = F + F_0$. By that we have only displaced the 'field strength' F by the constant F_0, the field energy as function of the new field G is shown in figure

7.6c, it is obtained from figure 7.6b by shifting the V-axis to the left, the formula for this curve is easily obtained by replacing in the above expression of V the variable F by $G + F_0$:

$$V = -2M \cdot G^2 - 4g \cdot \sqrt{\frac{-M}{2 \cdot g}} \cdot G^3 + g \cdot G^4 - \frac{M^2}{4 \cdot g} .$$

The expression now looks more complicated, in particular there appears now a term with G^3, a cubic term which destroys the symmetry under the transformations of F into $-F$. This is a consequence of spontaneous symmetry breaking: we have made a decision for $G = F + F_0$ but we could have decided equally well for $G = F - F_0$. Though it looks more complicated, this expression for V has a big advantage, the minimum of the field energy V occurs now for the field strength $G = 0$. This is reflected in the *positive* quadratic term $-2M \cdot G^2$ which allows now the following interpretation of the model: It describes a particle with mass squared equal $-2M$ and with a cubic (G^3) and a quartic (G^4) self interaction. This means we can view the model under two aspects: For symmetry considerations it is convenient to work with the field strength F, for particle interpretation in a quantum field theory it is more convenient to work with G.

In order to make the model really interesting, we have to extend it. The symmetry operation 'change of sign of F' is not a continuous operation, but we have seen that Goldstone bosons only occur if a *continuous* symmetry is spontaneously broken. Therefore we rotate the curves from Figure 7.6 a and b around the vertical (the V) axis. From curve (a) (with positive M) we obtain a bowl, Figure 7.7 a, from curve (b) a surface which is similar to the bottom of a champagne bottle, Figure 7.7 b. Expressed in mathematical formulæ, rotating the curves of figure 7.6 corresponds to the introduction of two fields, F_1 and F_2 with the static field energy

$$V(F_1, F_2) = M \cdot (F_1^2 + F_2^2) + \lambda \cdot (F_1^2 + F_2^2)^2 .$$

.

Now we have infinitely many ground states, namely the groove in the bottom of the champagne bottle. The field strengths of these ground states satisfy the equation:

$$(F_1^2 + F_2^2) = -\frac{M}{2 \cdot \lambda} .$$

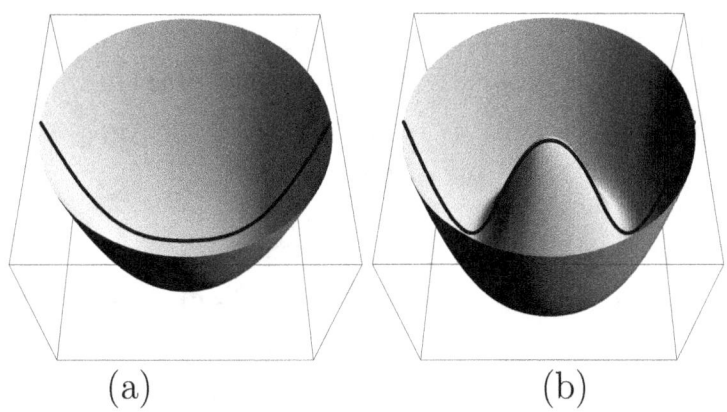

(a)   (b)

Figure 7.7: Static field energy $V(F_1, F_2)$ of a field with two components F_1 and F_2; (**a**) with positive quadratic term (mass term) ; (**b**) with negative quadratic term leading to a spontaneous symmetry breaking

The field energy and with that the dynamics is invariant under all transformations which leave $F_1^2 + F_2^2$ unchanged, this corresponds exactly to the rotation in the F_1, F_2-plane. In formulæ, this rotation is expressed by the substitution

$$F_1 \to \cos\theta \cdot F_1 + \sin\theta \cdot F_2; \qquad F_2 \to -\sin\theta \cdot F_1 + \cos\theta \cdot F_2 \,,$$

where the field strengths F_1 and F_2 depend on the space and time coordinates, but the 'rotation angle' θ is constant (that is we consider a global transformation).

It is very convenient to unite the two fields F_1 and F_2 to a complex field with F_1 as real and F_2 as imaginary part. Than the symmetry operation, the rotation in the F_1, F_2 plane, corresponds to a multiplication with a phase factor, this means it is a $U(1)$ transformation. We can thus summarize the situation: The model with a negative M-term shows a spontaneously broken $U(1)$ symmetry. It has infinitely many ground states, but if we want to describe a real system we have to choose a particular one, say $F_{10} = F_0$, $F_{20} = 0$, and the symmetry is broken by this particular choice.

Since in this case the symmetry is continuous, the theorem of Goldstone applies and there will be a field with massless field quanta. This one can see by substituting the fields F_1 and F_2 by new ones, as we have done in the case of one field previously. The calculation is not particularly complicated. One obtains besides the field with massless particles also a field with (positive) mass square $-2M$.

Spontaneous symmetry breaking reminds of the story of Buridians donkey, dear to medieval scholastics. This ass is said to have starved since he was positioned exactly between two hay stacks and could not decide which one to choose. We guess however that the donkey was well aware of the possibility of spontaneous symmetry breaking and decided *spontaneously* on one of them.

7.2 The Standard Model of Particle Physics

Phenomenology of elementary particles is since 1980 in an excellent state. Presumably there are more physicists who complain that the Standard Model is too good than those who complain that it is too bad. The reason for that paradoxical behaviour is that the excellent agreement between theory and experiment leaves little space for evidence of "new physics".

We shall shortly describe the development that lead to that model and then describe the Standard model of particle physics in a short and systematic way where heavy use is made of the formalism of symmetries, see chapter 7.1.

7.2.1 The Development of the Standard Model

At the turn of the 19th to the 20th century two developments of physics were evident. Firstly, the field theory of electric phenomena, as conceived by Faraday and put in its final mathematical form by Maxwell, could not be considered as a branch of mechanics in the sense Euler had developed mechanics of continua. Einstein even reversed the order: he took the symmetries of the Maxwell equations more serious than those of classical mechanics and he thereby modified the latter to relativistic mechanics. Secondly, around the same time there was evidence from statistical mechanics and atomic spectra that classical mechanics had to be modified essentially at the scale of atomic extensions, that is around a tenth of a nanometer. This first led to the 'old quantum mechanics' initiated by Planck in 1900 and essentially extended by Einstein and Bohr. The 'new quantum mechanics' was originated by Heisenberg in his paper on *On Quantum-theoretical Reinterpretation of Kinematical and Mechanical Relations*. Not even two month after this paper was submitted, Born and Jordan formulated Heisenbergs ideas in a systematic way and at the end of their paper they made "the attempt, to fit the laws of the electromagnetic field into the new theory". They introduced matrices, that is non-commuting operators, not only for the mechanical observables, but also for the electric and magnetic field. The next essential step towards a realistic quantum electrodynamics was due to Dirac (1927). He could already rely on the interpretation given in a sequel to the paper of Born and Jordan, the famous 'Dreimännerarbeit' (three-men paper) of 1925, where also Heisenberg participated. Dirac used his approach based on analogies of quantum theory with higher mechanics an introduced annihilation and creation operators for photons. Since he had the full dynamics incorporated in his approach, he could give a dynamical derivation of the famous relation between the spontaneous and the induced emission coefficient, established by Einstein in 1916/17. Dirac was emphasizing the particle character of the electromagnetic radiation (photons), but in the same year Jordan and Klein, following in some respect Dirac's ideas, stressed the opposite, namely the field character of matter. Jordan also realized that for fermion fields the commutation relations had to be substituted by anti-commutation relations.

Two papers authored by Heisenberg and Pauli and published in 1929 can be regarded as the first papers having the essential ingredients of relativistic quantum field theory. They treated both the matter fields and of course the radiation field relativistically. For the matter field they used the relativistic wave equation found by Dirac, which shall be mentioned later several times. They used the canonical formalism of classical field theory for the quantization procedure, in analogy to the application of the canonical formalism of mechanics in establishing quantum mechanics. On their way they met a tremendous obstacle: as a consequence of the Maxwell equations the conjugate field of the electric potential is zero. This and other difficulties made the two silent for nearly a year, a very long period in a time where seminal papers were often separated by only a few weeks. The real breakthrough came when they realized the importance of gauge invariance in quantum theory, a feature first clearly recognized by H. Weyl and already stressed in his famous book *Group Theory and Quantum Mechanics*, the first edition of which appeared in 1928. So the main ingredients of quantum field theory were found in a period of only four years an a few month after the first appearance of new quantum mechanics. But further progress was by no means easy. Heisenberg later remembers that in contrast to quantum mechanics quantum electrodynamics became never simple. The mood of the early 1930ies is caught in his reminiscence:

"In 23 and 24 we knew that there were difficulties and we also had the feeling that we were quite close to the final solution of the difficulties. . . . It was as if we were just before entering the harbor, while in this later period we were just going out into see again, i.e. all kinds of difficulies coming up . . . ."

we will not dwell on these difficulties mentioned. The outcome of the adventure on open sea was renormalized relativistic quantum field theory, which governed large parts of physics for the rest of the 20th century and is still going strong in the 21st.

Quantized field theory led in the sequel to a dichotomy with epistemological consequences. In the theoretical description the *field* concept is the fundamental one, but on the other hand all our knowledge comes from accelerated and detected *particles*. Only in perturbation theory there is a clear-cut relation between particles and fields: the field quanta are the (observed) particles.

Physics did not stop on the level of atoms. After the essential questions of atomic spectra had been clarified, nuclear physics entered the scene. The classical scattering experiments of Rutherford, Geiger and Marsden showed that the atoms had a nucleus which was extremely small as compared to the extension of the atom. The appropriate scale for the atom is the nanometer (10^{-9} m), that of the nucleus the femtometer (10^{-15} m). Elementary particles at the time were the electron and the proton, the nucleus of the Hydrogen atom. There were good reasons to believe that the nuclei of the other atoms were composite objects, its constituents being presumably protons and electrons. There was strong evidence for such a hypothesis: The mass of a nucleus was roughly an integer multiple of the mass of a proton and the charge was also a multiple of the charge of a proton, therefore the difference between the mass and charge number had to be explained by an extremely light negatively charged particle, just the typical properties of an electron. Furthermore the emission of electrons from a nucleus could be observed in the nuclear β-decay.

In β-decay there was, however, a serious problem. Chadwick and Ellis (1914-1927) had found that the electron spectrum in that decay was not discrete, as in the case of α-decay, but continuous. Furthermore there seemed for certain decays to be a problem with the relation between spin and statistics, if only one fermion was emitted. After Lise Meitner had, by her own experiment, convinced herself and Pauli of the correctness of the results of Chadwick and Ellis, Pauli found 'a desparate way out' from both problems: In β-decay not only an electron, but also another very light neutral fermion, later called neutrino, is emitted[2].

Besides this problems there were also some other serious difficulties to reconcile the otherwise good phenomenological evidence of the picture of the nucleus with theoretical principles. So it was not far fetched to assume 'new physics' to set in at the scale of the atomic nucleus, that is at several femtometer. Increasing the resolution by a factor of a million, that is from millimeter to nanometer, had led from classical to quantum physics, why should a further factor of a million, that is going from nanometer to femtometer, not also necessitate far reaching modifications. The scale where new physics should set in was generally considered to be the classical electron radius $r_e = \alpha\hbar/(m_e c) \approx 2.8$ fm.

The doubts that quantum physics could not be applied to scales much smaller than the atomic ones even influenced the interpretation of experiments. It was not clear that one could trust results obtained by quantum electrodynamics, for instance for the energy loss of charged particles in matter, when the wave length of the involved photons is of the order of a femtometer. It turned out soon, however, that such a transition to 'new physics' was not necessary and that quantum physics, as derived from atomic physics, also applied to nuclear physics. Several experimental and theoretical findings contributed to this insight. The α-decay of the nucleus was explained by Gamow (1928) as a quantum mechanical tunnel effect. Part of the theoretical problems of electrons inside a nucleus were solved through the discovery of the neutron by Chadwick 1932. It was immediately proposed (Heisenberg 1932) that the nucleus consisted of protons and neutrons rather than of protons and electrons.

In the same year local quantum field theory had its first spectacular triumph: the antiparticle of the electron, predicted by Dirac in 1928, was discovered in a cosmic ray experiment by Anderson. Though it was already predicted on the basis of local interaction in relativistic quantum mechanics, it is essentially a consequence of quantum field theory and can only be properly accounted for in a quantum-field-theoretical framework.

The neutrino hypothesis of Pauli was incorporated by Fermi in his

---

[2]This was communicated in an open letter to the 'radioactive ladies (L. Meitner was present) and gentlemen' at a meeting in December 1930

quantum field theoretical description of β-decay (1933). In this theory the occurrence of creation and annihilation operators for fermions was essential.

Though the interaction strength was very week, the theory had problems if one applied to it the procedures of perturbation theory used in quantum mechanics.[3] But on the other hand the lowest order (tree level) contributions of Fermi's theory were the basis for a very successful quantitative explanation of many observed decay spectra.

The success of quantum field theory in the description of β-decay, that is weak interactions, motivated Yukawa to develop a quantum field theory of nuclear forces (1935). In some sense it was closer to electrodynamics than to the Fermi's theory and it predicted as quantum of interaction the existence of a new kind of elementary particle, namely a massive particle with integer spin, which was first called mesotron, later π-meson. The mass (Compton wave length) should be corresponding to the size of nuclei, that is several hundred electron masses. A particle of such a mass was indeed discovered by Neddermayer and Anderson (1937), it turned out later, however, that it could not be the particle wanted for the Yukawa theory.

The discovery of the neutron had another very important impact on theory, it initiated the concept of internal symmetries. Since the mass of the neutron differs from that of the proton by only about 1 permill, Heisenberg proposed immediately a symmetry between the two particles, later called nucleons. On the basis of results of nuclear spectroscopy and first precise measurements of cross sections of proton-proton scattering, this theory was finally developed into the theory of isospin symmetry (Condon, Kemmer, Wigner and others).

The particle predicted by Yukawa, later called π-meson, was discovered in 1947 by Powell and collaborators, shortly after it had been shown that the mesotron, the particle found by Neddermayer and Anderson ten years earlier had not the properties to mediate strong interactions. The situation of particle physics seemed in the middle of the 20th century to be in a similarly good state as at the end of that century, though the Standard Model of that time was completely different from the present one. The elementary particles were: The proton, the neutron, the electron, the neutrino(s) and as particles me-

---

[3]These corrections were first derived in a truly mechanistic field theory, namely the theory of sound by Rayleigh 1877, in quantum mechanics they were derived by Max Born.

diating the electromagnetic and strong interaction the the photon and the π-meson, respectively. To that came a a particle, which 'nobody had odered' the muon, the former mesotron.

Quantum field theory turned out to be extremely successful. The problems occurring by just transposing the concepts of quantum mechanics to quantum field theory were solved by Dyson, Feynman, Gell-Mann, Schwinger and Tomunaga in renormalized perturbation theory of quantum electrodynamics and results were brilliantly confirmed by experiment (as they still are with increasing precision). The problem was that the prescription successful in nonrelativistic quantum mechanics led to infinite results. These difficulties were overcome in the renormalization procedure by ascribing the properties of the fundamental fields, like mass and interaction strength, no intrinsic value. The value choosen could well depend on the normalization procedure applied. This did not lead to ambiguities since it could be shown, that *observable* quantities do not depend on the special renormalization procedure chosen. This indicates that masses of the field quanta are not directly observable quantities. This seems astonishing since on can read in every textbook of particle physics values for mass of the electron, e.g.. The reason for that is that the electron can be observed as an isolated particle and there is a special renormalization procedure which assures that the mass occuring in the field equations for the electron coincides with the clasically measured mass of the isolated electron. This scheme is called the mass-shell enormalization scheme. We shall see that this plays for the fundamental fields of strong interactions, the so called quark fields, an essential role.

Quantum field theory was also the basis for a treatment of weak and strong interactions, though there were some flaws: In weak interactions the qualitative results were impressive, but the renormalization program, which was so successful in quantum electrodynamics, was not applicable without increasing the numbers of parameters indefinitely. In strong interactions, the problems were just the opposite. The field theory with pseudoscalar mesons was renormalizable, but the quantitative results of renormalized perturbation theory were by no means satisfactory. This was not unexpected, however, since the interaction constant between the nucleons and the π-mesons turned out to be several orders of magnitude larger than the electromagnetic coupling.

In contrast to today there were however strong signs that particle physics was more complex than the picture outlined above. This was inferred essentially from results of nuclear physics and confirmed by events produced by cosmic rays.

In 1947 Rochester and Butler discovered in a cloud chamber experiment traces with the topology of a V and which were called V-particles, today they are called *strange* particles. They were unstable but lived long enough to form traces in cloud chambers; their mass was definitely higher than that of a π-meson. Their unwanted presence could not be ignored by theoreticians for too long a time, especially since they were soon produced in large number in accelerator experiments. The development of accelerator and beam construction and of more and more refined detectors (e.g. bubble chambers) led soon to a true profusion of elementary particles which started a crisis for the whole field and initiated a search for new concepts. Since meson field theory did not lead to more than just qualitative results, G. Chew made the famous statement (1961):

> "we do not wish to assert (as does Landau) that conventional field theory is necessarily wrong, but only that it is sterile with respect to strong interactions and that, like an old soldier, it is destined not to die but just to fade away"

In weak interactions there was, from a strictly phenomenological point of view, no need to look for new concepts. Experimentalists were looking for the field quantum of weak interactions, the so called intermediate boson, but even if the search had been successful, the presence of an intermediate boson alone would not have solved the theoretical problem of non-renormalizability of the Fermi theory. Furthermore it was not clear if non-renormalizability was only a problem of weak interactions, since it was not known how strong interactions can influence weak interactions at small distances.

In strong interactions several lines of research, partially in parallel, partially in contradiction to each other were followed. All of them were motivated and inspired by quantum field theory, but none of them was willing to accept its full program, namely to calculate observable quantities directly from a Lagrangian. They all tried to handle the problem of the ever increasing number of *elementary* particles:

1. In the theory of the analytic S-matrix one tried to eliminate the field concept from strong interactions and concentrate on properties of scattering matrix elements derived solely from conservation of proba-

bility. Though part of this program had a strong affect against field theory (see the quotation of Chew from above), many of the postulated analytic properties of the S-matrix were results obtained in the framework of local quantum field theory. The approach culminated in the concept of 'nuclear democracy', in which all observed strongly interacting particles and resonances were treated on the same footing and were related through self-consistency conditions. This was the so called *bootstrap program*. The application of Regge's theory of potential scattering to high energy scattering and the use of dispersion relations in particle physics were an outcome of this program. Another important consequence of the theory was a model developed by Veneziano: It showed duality, that is it related the high energy behaviour of the scattering-matrix elements to the resonance structure (poles) of the matrix elements. It eventually gave rise to string theory.

2. There was a strong emphasis on internal symmetries, motivated by the success of the isospin symmetry $SU(2)$ in the analysis of π-meson-nucleon scattering.

3. The discovery of 'several new particles' led already 1949 Fermi and Yang to speculate that not all of them were elementary. They therefore proposed, rather as an illustration of a possible program than as a realistic model, to consider the π-meson as a bound state of a nucleon and an anti-nucleon. Though Fermi was coauthor of this paper, the idea was not enthusiastically embraced by the majority of the community. But the phenomenological evidence for the composite nature of strongly interacting particles grew with time. In figure 7.8 the hydrogen spectrum is compared with the spectrum of the nucleons, that is the particles and resonances with baryon number 1 and isospin 1/2. The search for a constituent picture of the strongly interacting particles led eventually to the phenomenologically very successful quark model of Gell-Mann and Zweig.

From the concepts mentioned above only the bootstrap philosophy has disappeared. Regge theory is a prerequisite for the description of hadronic high energy scattering processes and it gave birth, through the Veneziano model, to string theory. Dispersion relations are not in the focus of present-day theoretical interest, but they are still an important tool in the analysis of strong interactions. The second and third point are cornerstones of the present Standard Model, but it was a long and tedious way to incorporate these concepts into the frame of relativistic quantum field theory.

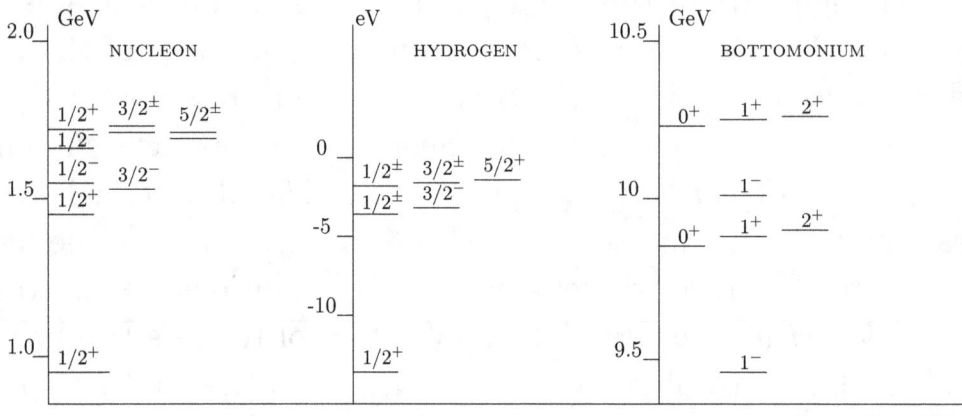

Figure 7.8: Lowest lying states of the spectra of the nucleon, hydrogen and the heavy meson state bottomonium

Before we come to a description of this Standard Model, we shall just quote a few important steps which lead to his final establishment.

An important step, though not recognized immediately as such, was the construction of a classical gauge field theory in which the gauge field transform under a non-Abelian symmetry group (Yang and Mills theory, 1954). It took however some time before this theory was formulated as a quantum field theory in the sense of a formal power series, or non-perturbatively, on a discrete set of space and time points (lattice).

In 1967 a theory of weak and electromagnetic interactions was proposed based on a classical Lagrangian, gauge invariant under $SU(2) \times U(1)$, with a mechanism for mass generation of the inter-action quanta (massive gauge bosons). It led to the prediction of neutral weak currents, that is to reactions like $\bar{\nu}_\mu + e \to \bar{\nu}_\mu + e$. It also led, together with the experimentally confirmed absence of strangeness changing neutral currents, to the prediction of a new quantum number, besides isospin and strangeness, later called *charm* (GIM mechanism, after its inventors Glashow, Iliopoulos and Mainai). 1971 the proof of renormalizability of the interaction based on the classical $SU(2) \times U(1)$ (electroweak) Lagrangian was finished ('t Hooft and Veltman).

Though the theory was now in a good shape, it was evidently not taken too seriously in the community. The search for neutral currents was only on position 8 in a priority list of 10 points of the relevant Gargamelle experiment. However the experimentalists, who in an heroic effort found 1972 three events of $\bar{\nu}_\mu + e$ scattering in

268

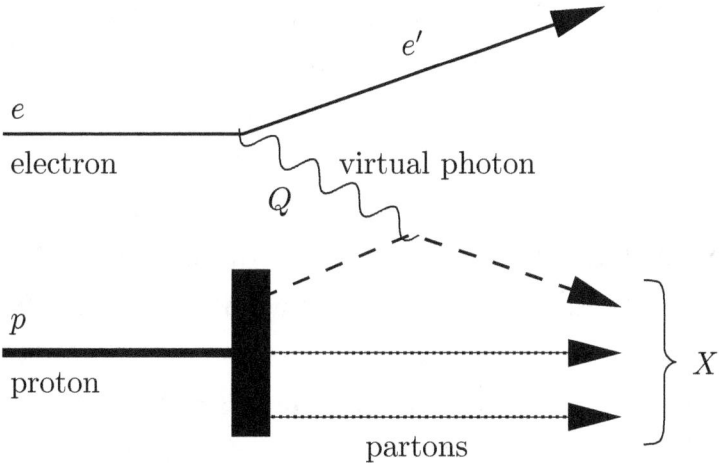

Figure 7.9: Schematic description of the parton model in deep inelastic scattering. The virtual photon interacts with only one parton, the other ones are not effected. This intuitive picture makes only sense in a reference frame where the momentum of the proton tends to infinity

1.4 Million pictures, write that they were motivated by the proof of renormalisability of the electroweak Lagrangian. This discovery of neutral currents opened the way for the general acceptance of the electroweak $SU(2) \times U(1)$ model. The Nobel price was awarded to Glashow, Salam and Weinberg for their contribution to the theory of unification of weak and electromagnetic interactions in 1979, before the quanta of the weak interaction, the massive gauge bosons, were found experimentally in 1983 at CERN.

The development of the gauge theory of strong interactions was a bit slower and there the interplay between experiment and theory was even stronger.

In the deep inelastic scattering experiments at SLAC (1966 ff) electrons with high energy were scattered off protons and especially reactions with high momentum transfer (more than 1 GeV^2) were analyzed. It turned out that special features of these reactions were best described by a picture in which the proton consisted of a bunch of practically free constituents, the so called partons. A scheme of this picture is given in figure 7.9

The detection of the heavy J/ψ-(1974) and Υ-(1976) meson and their resonances made it even more evident that a bound state picture of hadrons could explain many features (see Figure 1.8,last row: bottomonium).

In 1973 Fritzsch, Gell-Mann, and Leutwyler proposed quantum

chromodynamics (QCD) as the dynamical theory of strong interactions. It was a gauge theory based on unbroken $SU(3)$ (colour) symmetry. Its phenomenological basis was the success of two different approaches, namely current algebra and the quark model. It could explain extremely well the deep inelastic scattering experiments and, with some extra ingredients, the spectra of the J/ψ- and Υ−states.

An essential ingredient of the electroweak Lagrangian, the so called Higgs boson, has now been detected at the Large Hadron Collider in CERN, see section 7.2.6. The only very clear-cut evidences that the Standard Model has to be modified in its present form are the neutrino oscillations which in the most favourable case would lead to a rather straightforward extension with 9 new parameters. Apart from this major and some minor *black clouds* there is a very nice blue sky over the model. In the next section it will be described more systematically.

7.2.2 Systematic Description of the Standard Model

We shall describe the Standard Model in the form it had before the discovery of neutrino oscillations, since the modifications due to this phenomenon are not yet clear. They will however change very little of the present-day phenomenology, though they might be very important for the future development.

We have mentioned in the preceding section that the relation between field and particle concept is far from being trivial and depends to some extend on the theoretical frame. Nevertheless we shall often adopt here the usual jargon and speak of particles even when only the field theoretical aspect is fully consistent.

The Standard Model of particle physics is based on 3 families of fundamental fermion quantum fields (i.e. spin $\frac{1}{2}$, see sect. 7.1.1). Each family contains two fields of fractional elementary charge[4], the so called quark fields and of two fields called lepton fields. One of the lepton fields of the first family has as field quantum the electron, a particle known since 1897. It has negative elementary charge, the other particle is the so called neutrino, a (nearly) massless particle which carries no charge and does not participate in strong and electromagnetic interactions. It only participates in weak interactions and, as all matter does, in gravitational interaction. The quark and

---

[4]The charge of the proton (nucleus of the hydrogen atom) is the unit charge in particle physics

electron fields are composed of a right and left handed spinor field, that is the direction of the spin can point as well in the direction of motion as well as opposite to it (see sect. 7.1.1). The neutrino fields are left handed, that means the spin of particles can only point in the direction opposite to the direction of motion.

Strong Interactions, QCD

In strong interactions, only the quark fields participate. Each quark field occurs, beside the two spin directions, in three states, called *colors*. These colors transform according to a gauged (see 7.1.2) $SU(3)$ group which fixes the form of the interaction completely, only the strength is not determined by the gauge symmetry. Since $SU(3)$ has three generators, there are 8 quantum fields with spin 1 which mediate the strong interaction, these are the gluon fields. The interaction theory of quarks and gluons, that is the fundamental theory of strong interactions, is called quantum chromodynamics (QCD).

To the quanta of the quark fields there correspond no observable particles, therefore there is no mass-shell renormalization scheme for them: the quark mass inherently depends on the renormalization procedure. The strongly interacting particles which are observable, protons or neutrons e.g., are composed of quark fields. By far not all features of strong interactions can be calculated from the fundamental interaction QCD, but all those which could be calculated are in agreement with experiment and many more properties can be approximately calculated in models based on QCD.

There were several very important predictions made by QCD. One was the existence of so called three jet events. If an electron and its antiparticle, the positron, annihilate they can create strongly interacting particles and antiparticles like (anti-)protons and mesons. If the gluon is indeed a fundamental field in strong interactions, then one should observe so called three-jet-events, that is events were the particles produced in the annihilation process form three streams (jets) going in different space directions. One stream is due to a produced gluon, the two others to a quark and antiquark. Two jet events had been observed for some time, they are due to the production of a quark and an antiquark, the observation of three-jet-events (see figure 7.11) at the high energy laboratory DESY at Hamburg is considered in a somewhat oversimplified manner as the experimental proof for

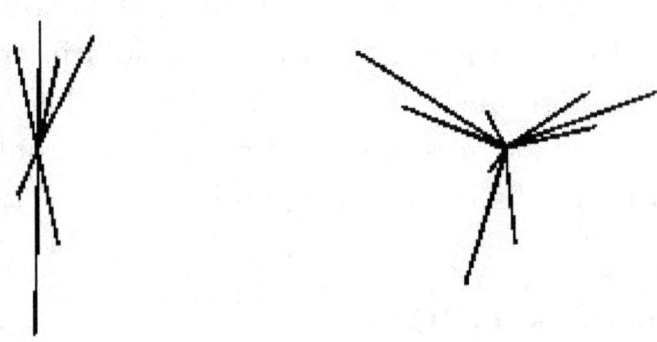

Figure 7.10: Schematical picture of two jet events (left) and the the observed three-jet event of DESY (right); the lines are traces of hadrons, the jet axes are clearly visible. Three jet events are clear indications for the gluon field, the gauge field of strong interactions (QCD)

the existence of the gluon (see Söding (2010)).

A special feature of QCD is asymptotic freedom: the coupling strength decreases with decreasing distance. Therefore for processes where only small distances play an important role the highly developed machinery of renormalized perturbation theory can be applied also in strong interactions. This led to many very well confirmed theoretical predictions. Especially important were precise predictions of the behaviour od scattering of electrons on protons, where a high fraction of momentum is transferred from the electron to the proton. High momentum transfer (deep inelastic scattering). As mentioned above in a particularly simple model, the constituents of the proton were assumed practically not to interact among each other (see Figure 7.9). This has as a consequence, that the so called structure function, the essential information on electron-proton scattering is independent of the momentum transfer, but depends only on the ratio of momentum transfer to energy. In QCD this simple picture is modified: the quarks constituting the proton interact, but if the momentum transfer is strong enough asymptotic freedom makes the interaction so small, that the dependence on the momentum transfer, not present in the simple model, can be calculated in perturbation theory. In figure 7.11 the structure function for two different values of momentum transfer is shown. In the simple model, the points should coincide for all values of x, the ratio of momentum transfer to energy. The curve for the squared momentum transfer $Q^2 = 90$ GeV^2 has been calculated from the curve describing the points for $Q^2 = 3.5$ GeV^2 using QCD.

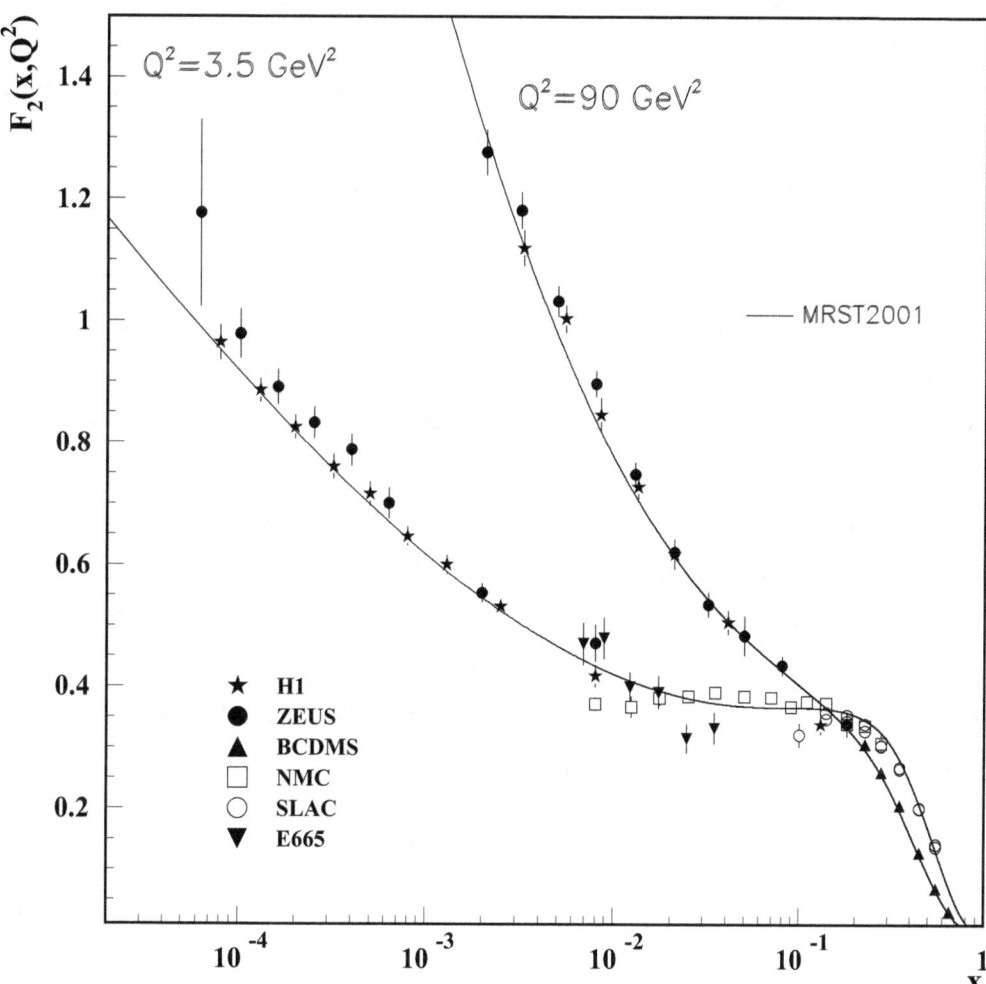

Figure 7.11: Experimental points and theoretical curves for the structure function $F_2(x, Q^2)$ as function of x, the ratio of momentum transfer to energy for two different values of the squared momentum transfer Q^2. The theoretical curves would coincide in the simple noninteracting model (which is indeed nearly the case for values near $x = 1$). The curve for displayed for $Q^2 = 90$ GeV^2 has been calculated from the curve describing the points for $Q^2 = 3.5$ GeV^2 using QCD.

The isospin symmetry, the first observed internal symmetry which played such an important role in the development of physics is in the frame of QCD rather accidental. it is due to the similarity of the masses of the two quarks of the first family. It could turn however that it has deep roots, if one understands better the mechanism of mass generation, which will be discussed below.

There is, however, still a big open question in QCD. Though the quark and gluon fields play the essential role in the theory – the laws of the theory are described in their terms – we cannot derive analytically from the fundamental laws of QCD, why the quarks and gluons cannot be observed as isolated particles. The neutrons and protons are also bound inside the atomic nucleus by strong forces, but nevertheless they can be experimentally isolated by smashing the nucleus. This is not the case for quarks and gluons and that is a completely new situation in physics [5]. One can read in any popular and not so popular book on particle physics, that protons and neutrons consist of quarks and gluons, but they cannot be separated into these particles. It is a bit as if Lavoisier had said: Water consists of the elements Hydrogen and Oxygen, but one cannot separate water into these elements. It is doubtful that the new chemistry had found many followers under these circumstances. But in QCD this is exactly the case, and one has invented a nice word for this special feature, that the elementary fields cannot be isolated: *confinement*.

In order to end this subsection on not too pessimistic a note, one should remark that numerical simulations[6] indicate very strongly, that this feature of confinement is indeed a consequence of its laws, but for one, who finds an analytical proof of it, there is still a price of one million dollar waiting.

Electromagnetic and Weak Interactions

Electromagnetic Interaction. After gravity the electromagnetic interaction is the one longest known to mankind. Considered some time rather as a curiosity, it urned out to be the one which affects perhaps most our daily live: chemistry, emission of light, all hardware of in-

---

[5] For the delicate relation between particles and fields in QFT see Dosch (2011)

[6] In these simulations the continuous space time is approximated by a lattice with a small, but finite distance between the lattice points. Even in these numerical simulations in quantum field theory the gauge principle plays an essential role. Progress in numerical calculations was not only due to the powerful supercomputers, but also due to the fact, that one had learned how to transpose the gauge principle from continuous space-time to a discrete lattice (Wegner 1971, Wilson 1974)

formation technology is governed by electromagnetic interaction. The laws of electromagnetic interaction were brought by J. Clerk Maxwell in the 1850 into the form which still are the basis of the modern theory[7].

In the formalism shortly sketched in sect. 7.1.2 the whole electrodynamics can be very concisely summarized as:

- electrodynamics is the gauge theory of the group $U(1)$ where the matter fields are the quark and electron fields, the gauge field is the electromagnetic field, its field quantum the photon.

Weak Interaction. In weak interactions the situation is more complicated. The gauge group is here a product of two groups, namely a $SU(2)$ and $SU(1)$, that is this groups contains as well the group $SU(2)$ as the group $U(1)$. The formation of such a product group $SU(2) \otimes U(1)$ is mathematically well defined. The fundamental doublet which transforms under the $SU(2)$ part is formed of *lefthanded* particles alone, the singlet of the $U(1)$ part is a right handed particle.

For the leptons the lefthanded doublet is formed of the charged lepton and its neutrino, the singlet of the right handed charged lepton. For quarks the situation is more complicated, since the families mix. If this where the whole situation, the gauge bosons of the weak theory, three for the $SU(2)$ part and one for the $U(1)$, would be massless. But the situation is complicated considerably since there has to be introduced an additional matter field, the so called Higgs Boson. This Higgs boson develops through spontaneous symmetry breaking a finite vacuum expectation value, that is the state of minimal energy is not reached if the field value of the Higgs field vanishes. This vacuum expectation value in turn leads to a mass of the gauge bosons of weak interaction. This Higgs boson can also give a mass to the leptons and the quarks.

There is an additional complication in the sector of weak and electomagnetic interactions: they mix in the $U(1)$ part, so we have in reality an electroweak interaction.

The Electroweak Interaction. The gauge bosons of the $U(1)$ of electrodynamics and of weak interaction have the same quantum

---

[7]It is remarkable that only the knowledge of these precise laws allowed the second industrial revolution. It started with the application of Maxwells equations to the construction of radio wave emitters and these are – together with quantum physics – still the basis of all information technology

numbers and hence they can mix. The result of this mixture, parameterized by the electroweak mixing angle θ_W, are two gauge bosons with different mass: a massless one, which is the photon, and a rather heavy one, which is called Z^0 boson. The mixing between the weak and electromagnetic interaction is however no true unification, since there remain still two independent gauge couplings.

An important consequence of the existence of neutral gauge bosons in weak interactions is the existence of so called weak neutral currents. Before they were discovered consistency conditions led to the prediction of a new quark, the later called charm quark. The detection of neutral currents was a major experimental effort. 100 000 bubble chamber pictures had to be scanned, before the first definite reaction induced by a neutral currents has been found.

The Parameters of the Standard Model

The elementary fields of the Standard model are collected in table 7.1.

The parameters of the Standard model are, not taking into account neutrino mixing:
• 3 gauge gauge couplings, one for the colour gauge group $SU(3)$ of strong interactions, one for the $SU(2)$ part and one for the $U(1)$ part of the electroweak gauge group
• 9 mass parameters for the three charged leptons and 6 quarks
• 4 parameters in the CKM quark mixing matrix
• 2 parameters in the Higgs part of the Lagrangian: The self coupling λ and the vacuum expectation value v of the Higgs doublet
To this one may add a 19th parameter:
• the so called θ parameter[8].

Incorporation of neutrino mixing will at least add 9 new free parameters.

The importance of the Standard model can also be seen from the 16 Nobel prices (10 experimental, 6 theoretical) awarded for cintributions directly related to the development of the Standard model, the latest 2008 for spontaneous symmetry breaking and qurk mixin in weak interactions.

---

[8]The θ term is an additional term in the pure gauge part of the Lagrangian which does not change the equations of motion but can lead to CP-violation

The elementary fields of the Standard Model

Field	spin	charge	mass	width
leptons				
electron	1/2	-1	0.5109989 MeV$/c^2$	stable
muon	1/2	-1	105.63806 MeV$/c^2$	$3 \cdot 10^{-16}$ MeV$/c^2$
τ-lepton	1/2	-1	1776.99 ± 0.29 MeV$/c^2$	$2.3 \cdot 10^{-9}$ MeV$/c^2$
e-neutrino	1/2	0	< 3 eV$/c^2$	
μ-neutrino	1/2	0	< 190 eV$/c^2$	
τ-neutrino	1/2	0	< 18.2 MeV$/c^2$	
gauge bosons of electroweak interactions				
photon	1	0	0	stable
$W^{\pm}$-boson	1	± 1	80.425 ± 0.039 GeV$/c^2$	2.118 ± 0.042 GeV$/c^2$
Z-boson	1	0	91.1876 ± 0.0021 GeV$/c^2$	2.4952 ± 0.0023 GeV$/c^2$
Higgs-boson				
H^0	0	0	125.6 GeV$/c^2$	
quarks				
d-quark	1/2	$-1/3$	5 to 8.5 MeV$/c^2$	
u-quark	1/2	$2/3$	1.5 to 4.5 MeV$/c^2$	stable
s-quark	1/2	$-1/3$	80 to 155 MeV$/c^2$	
c-quark	1/2	$2/3$	1.0 to 1.4 GeV$/c^2$	
b-quark	1/2	$-1/3$	4 to 4.5 GeV$/c^2$	
t-quark	1/2	$2/3$	174.4 ± 5 GeV$/c^2$	
gauge boson of strong interactions				
gluon	1	0	0	

Table 7.1: The charges are those of the particles, antiparticles have opposite charge. The masses of the quarks are parameters which make only sense in a definite renormalization.

7.2.3 Mass Generation

In the simplest of all possible worlds all particles would be mass-less. The fermions, that is leptons and quarks, would be described by Weyl-spinors (see 7.1.1. These obey the chiral symmetry which excludes a mass. The gauge bosons, that is the photons, the intermediate $W^{\pm}$- and Z^0-bosons, and the gluons would all be massless because auf the gauge symmetry. But we have seen that through the Higgs-boson and the spontaneous symmetry breaking the electroweak gauge bosons as well as the fermions acquire a mass. Nevertheless the statement "all particles are massless" is true to a very good approximation, namely if we use as unit of mass the only inherent scale in fundamental physics, the Plank mass. It is obtained if one combines the following fundamental constants of nature: Newton's gravitational constant gG_N, the velocity of light c, and Plancks constant $\hbar$. Its value, $\sqrt{\hbar c/G_N} = 1.22 \cdot 10^{19}$ GeV/c^2 is indeed unimaginably large compared to the masses of the fundamental particles listed in Table 7.1, that the above quoted approximation is indeed excellent.

In this light we can regard the families of fermions and the gauge bosons as massless to a very good approximation and the above mentioned symmetries give us a good reason fir that. Rests one particle, the Higgs-boson. For that particle none of the known symmetries predicts a vanishing mass value, and therefore its natural mass would be of the order of the Planck mass. The Higgs boson, or more precisely a particle which up to now shows all the features that one expects from this particle has a mass of 125.6 GeV/c^2. This is far, far below the Planck mass. There should be an additional symmetry which makes also the Higgs-boson light. Such a symmetry is the so called supersymmetry discussed in section which we shall very shortly discuss later.

In the last years a possibility to obtain a smaller value for the natural mass unit has been discussed intensively: It is based on the fact that general relativity is very well tested for large distances but not for small ones. It is possible that Newtons law, according to which the gravitational force falls of with the inverse square of distance, is not fulfilled for very small distances. Such deviations from Newton's law can be due to higher dimensions. In such theories our three dimensional space is embedded into a space of higher dimensions, which can only be penetrated by gravitational forces. In such a higher dimensional space the Planck mass would be much smaller. Two additional space dimensions would reduce it to approximately 100 GeV/c^2, that is exactly the range of electroweak symmetry breaking.

Theorie with extra space dimensions had been discussed earlier by T. KaluzaKaluza, T. (1921) and O. KleinKlein, O. (1926); they were stimulated by the work of Weyl on gauge symmetry in gravitational theory. Since even the most ivory-towered mathematical physicist has to move in three space and one time dimension, he must admit that the extra dimensions cannot be treated on the same footing as the familiar ones. The problem is solved by *compactification* of the extra dimensions. Such a compactification can be viewed very intuitively: A sheet of paper is a two-dimensional object, but if it is rolled up very tightly, than one of its dimensions, its length, is still visible "macroscopically", but viewed from a large distance on dimension seems to have disappeared, it has been compactified.

The hierarchy of the masses of the fundamental particles of the Standard model is an ambitious but unsolved problem in particle physics.

7.2.4   Some Speculations Concerning the Near Future

Grand Unified Theories (GUT)

A reproach very often made to the Standard model is that it contains too many parameters for a fundamental theory. Also the fact that there are three families of quarks and leptons which are unrelated is somewhat repugnant to the quest for unification of fundamental physics. An obvious answer to these objections is to embed the gauge groups of the Standard Model, that is $SU(3) \times SU(2) \times U(1)$ into a simple group, for instance $SU(5)$. In that case we have only one gauge coupling. Such a large symmetry is called a GUT (from Grand Unified Theory) symmetry. This symmetry is supposed to be broken at a certain scale M_G, and all particles which do not occur in the Standard Model have masses larger than that scale. Since this scale is supposed to be very high, there is little hope to observe these new particles directly. The Standard Model is an effective theory of the GUT, this means that in the model all degrees of freedom relevant at scales higher than M_G are absorbed in the renormalized parameters of the Standard Model. This is possible through the decoupling of the heavy particles – even inside loops – if the external scales are small compared to M_G and the theory is renormalizable. Renormalization group arguments based on the gauge group of the Standard Model may be applied and yield the observed, experimentally well established scaling behaviour for the strong and the electroweak gauge couplings. If the scale comes, however, into the region of the GUT scale M_G, the heavy states can no longer be neglected and the different gauge couplings of the Standard Model must meet and follow together the renormalization group equation of the unified gauge group, e.g. $SU(5)$. In this way the grand unification scale M_G is constrained by the low energy effective theory. Furthermore there is a stringent consistency condition for the effective theory since all three gauge couplings have to meet at the same point.

The GUT theory will generally lead to additional interactions in the Standard Model which from the point of view of the effective theory might not be renormalizable, but they are suppressed by powers of μ/M_G, where μ is a scale typical for the effective theory, that is $\mu \ll M_G$. A nice example of an effective theory is the Fermi theory of weak interactions. It is an effective theory of the Standard Model at scales small compared to the mass of the gauge bosons W and Z. In it occurs the unrenormalizable four-fermion coupling, but the coupling constant G_F in front of it is proportional to M_W^{-2},

$$G_F = \frac{\sqrt{2}g^2}{8M_W^2}$$

Important additional terms introduced by GUT are interactions leading to the decay of the proton into leptons and other non-baryonic states. These effects will be small since they are suppressed by powers of M_G. Nevertheless experimental limits on the proton decay in specific channels have already falsified a lot of proposals for grand unified theories. Present limits for important decay channels are, with 90% confidence level:

$$t_{p \to e^+ \pi^0} \geq 1.6 \cdot 10^{33} \text{ years} \quad t_{n \to e^+ \pi^-} \geq 0.16 \cdot 10^{33} \text{ years}$$

The Supersymmetric Grand Unified Standard Model

At the moment the most widely accepted extension of the Standard Model is a supersymmetric GUT, that is a gauge theory which has at low energy – low with respect to M_G – is the supersymmetric Standard Model where the supersymmetry is broken at a scale of around 1 TeV.

The concept of a supersymmetry is an extension of the concept of symmetry by adding a new kind of generators (see sect 7.1.1) to those of the group. The effect these extension is that fields with different spin, those with spin $\frac{1}{2}$ and spin 1 are related through supersymmetric transformations.

The indications that this might be indeed the next Standard Model are:

1. The supersymmetric Standard Model predicts a light Higgs, $m_H < 170$ GeV/c^2. This is a value well inside the range of theoretical predictions based on precision experiments.

2. As mentioned in 7.2.2 the coupling constants in gauge theories depend on the choosen renormalization scale. If a (broken) supersymmetric model the strong and the electromagnetic and the weak gauge coupling fuse to become an unique value at a scale of about 10^{16} GeV. Such a fusion to only one constant is of course necessary for a Grand Unified Theory.

3. This above mentioned scale can lead to a proton lifetime compatible with present day experimental bounds (but close to them).

The supersymmetric Standard Model is still a speculative model since up to now no particle predicted by supersymmetry has been observed, but the speculations are based on present day experimental findings; they make also stringent predictions for the near future:

1. The light Higgs-boson should be observed at he new accelerator LHC at CERN, Geneva.

2. The proton decay should be observed in the near future

3. Supersymmetric partners of known particles should most probably seen at LHC

If these predictions will be fulfilled, then the supersymmetric GUT will be the new Standard Model. Precision experiments could then indirectly explore the large desert from the TeV region to 10^{16} GeV.

hier Nachtrag

7.2.5 Holographic QCD, also an Epistemological Challenge

We have several times mentioned string theory, especially in sect. 4.62 and in this chapter. Though it has its origin in the theory of strong interactions it turned out to be one of the most exotic theories of physics. Not only is its space-time high dimensional, more than 10 dimensions are necessary for a consistent description, but also the typical feature of a string, namely its excitations are of extremely high energy. With present day technique one would need an accelerator with the radius larger than the orbit of the planet Neptun to excite the lowest mode. Nevertheless this theory, or more precisely a small segment of string theory, has excited also very down to earth physicists, which are primarily interested in phenomena. This is a consequence of a well founded conjecture, the so called Maldacena conjecture (Maldacena 1998, Witten 1998, Gubser et al 1998). This conjecture states that a 5-dimensional subspace of a certain string theory in ten dimensions corresponds to a gauge theory, that is a theory with quarks and gluons, in 4 space-time dimensions. Though this gauge theory is described in the conventional number of dimensions, 3 for space, 1 for time, it is far away from the QCD of the Standard model: It shows a very high supersymmetry which not observed. Nevertheless it could be extremely interesting for QCD. Under certain circumstances the correspondence is dual, that is processes with strong interactions in the gauge theory (supersymmetric QCD) correspond to processes with weak couplings. Confinement of quarks and gluons (see sect. 7.2.2), i.e. the feature that the latter cannot be produced as isolated particles, is certainly a process where strong interactions play an essential role, the interaction is so strong that you cannot separate the alleged "constituents". Therefor on hopes that in the dual theory, the one in 5 dimensions, one can derive analytically results not obtainable in QCD. In a so called bottom-up approach one starts with QCD and looks for its 5-dimensional dual partner. This is called holographic QCD, in analogy to optical holography where the two dimensional hologram contains **all** the information about objects in 3 dimensional space.

Independent of an eventual success of this program, already the Maldacena conjecture has important consequences for scientific realism. We have two theories: one dealing of quarks and gluons which are the unobservable "constituents" of observable particles (hadrons in real QCD) and one dealing only with the fields of the observable particles, but accommodated in a 5-dimensional space. If these two theories are equivalent, the two questions:" Do quarks and gluons exist?" and" Is the fifth dimension real?" are on the same footing.

In this way the Maldacena conjecture is a beautiful – up to now only theoretical – illustration of Hertz's (1894, p. 1ff) theory of sign's and symbols. He states explicitly that our theories are not a mirror of reality, but simulacra (Scheinbilder) or symbols, which have to fulfill certain axioms. But he notes explicitly, that these symbols have not to be unique and their choice is often a question of intellectual economy. This ambiguity might be more palatable if one considers a theory of complex systems than for the case of "the last constituents of matter". In the epilogue we come shortly back to this question.

7.2.6 Note added in proof: Some Recent Discoveries

On July, 4th, 2012, the two collaborations CMS and ATLAS, both working at the Large Hadron Collider (LHC) at CERN, could announce that the Higgs boson, (see sect. 7.2.2), Electromagnetic and Weak Interactions) has been finally detected and its mass was found to be ca 126 GeV/c^2. As in most experiments, there is always a certain probability that the effects one looks for is accidentally created by "background events". This is a particularly serious problem in high energy experiments, where thousands of particles are created and can simulate many effects. Therefore in physics one has the convention that an observation is called "effect", when the probability that it is created by accidental configurations is less than 3/1000 (3 σ deviation), and it is called a discovery, when the probability is less than 1/1 000 000, (5 σ deviation)[9]. This level of significance has been reached at the LHC independently by both groups and was presented in a seminar at CERN on July, 4th 2012. To this statistical certainty it must be added that in the Standard model of particle physics a particle with these properties in that mass range had been predicted for some time.

The Higgs particle was proposed in 1964 independently by P. Higgs, and F. Englert and R. Brout, and G.S. Guralnik, C.R. Hagen and T.W. Kibble. It was introduced to give the meson fields, which mediate weak interaction, a mass (see sect. 7.1.3). It contributes, however, very little to the mass of ordinary matter. How this mass is created is closely related to the confinement problem in QCD (see 7.2.2, Strong Interactions, QCD).

In the popular press the Higgs boson is sometimes called "The God particle". Of course no serious physicist would use this name and its

---

[9]In life sciences an effect is called significant, if the level of security is 95 %, corresponding to 2 σ

origin is rather typical for popularization of science. The Nobel laureate L. Lederman called it "The goddam particle", because it was for a long time so elusive. But his editor changed the name, presumably for commercial reasons, to "The God particle".

The expectations to find indications for supersymmetry (see 7.2.4, The Supersymmetric Grand Unified Standard Model), however, have not been fulfilled, at least up to now.

There was another (short) excitement in connection with the LHC. In September 2011 the OPERA group, working at the Gran Sasso laboratory in Italy, had announced that the neutrinos travelled faster than light from Geneva to the Gran Sasso tunnel (near Rome). This result was theoretically totally unexpected and would essentially modify one of the best established symmetries in physics, the Lorentz symmetry (see sect. 7.1.1), which is the heart of the theory of special relativity. Though the results appeared to be statistically highly significant, most physicist were very sceptical. If normal particles could indeed travel (a tiny bit) faster than light in the vacuum, it would have reversed essential parts of the core-context. The scepticism was justified: in February 2012 it went public that there were serious flaws in the experiments and the result was withdrawn.

7.3   The Standard Model of Cosmology (SMC)

The Standard Model of Cosmology[10] is principally based on two well founded hypotheses, namely

- on the validity of Einstein's equations of general relativity, and

- on the hypothesis that on a sufficiently large scale matter is homogeneously distributed.(Cosmological hypothesis)

Einstein's equations have been brilliantly corroborated in many instances and there is also no observational evidence which contradicts the cosmological hypothesis[11]

Under the cosmological hypothesis Einstein's equations are drastically simplified: The 10 coupled partial differential equations for the general case reduce to two single normal differential equations, the

---

[10]For an excellent non-technical treatise see Damour(2006)

[11]of course as ordinary matter is concerned, the averaging has to be done over sufficiently large volumina which contain several galactic clusters.

so called Friedmann-Lemaître equations. These equations contain as input, besides the Newtonian gravitational constant G_N, the (by hypothesis homogeneous) density of the universe, the pressure of the universe and a cosmological constant. The Einstein equations determine in general the four dimensional space time geometry of the universe. Under the cosmological hypothesis geometry simplifies to the so called Robertson-Walker metric. It contains only two free parameters, the unit length R and a parameter k responsible for the overall curvature of the universe. This parameter k can assume three values, -1, 0 or $+1$; for $k = -1$ the universe is open, for $k = +1$ it is closed and for $k = 0$ it is asymptotically flat, in the last case which seems to be realized in our universe, Euclidean geometry holds on the large scale. The equations are so simple, that one can even understand them with high school mathematics: If the unit scale is denoted by R, its rate of change with time by $\dot{R}$ and its acceleration by $\ddot{R}$ the equations are:

$$\left(\dot{R}/R\right)^2 = \tfrac{8}{3}\pi G_N \rho - k/R^2 + \tfrac{1}{3}\Lambda \quad \text{and} \quad \ddot{R}/R = \Lambda/3 - \tfrac{4}{3}\pi G_N(\rho + 3p)$$

where ρ is the density of the universe, p the pressure and Λ is the so called cosmological constant and k the parameter which determines geometry.

Though four-dimensional geometry goes beyond our direct intuition, it is useful to consider the two cases for two dimensions. In that case, the surface of a sphere forms a closed space, with the inverse radius of the sphere as curvature, a plane with Euclidean geometry corresponds to a flat space and the surface of a hyperboloid to that of an open space, see Figure 7.12.

The quantity to be determined by the equations is that of the unit length depending on the only nontrivial variable left, that of time. This unit scale is a measurable quantity. We can for instance take the wavelength of a certain spectral line as its measure. A change of the unit length with time can be observed as a change of the wavelength of that characteristic spectral line. To give an example: The helium in the sun emits (among other lines) light with the characteristic wave length of 0.00059 mm (this wavelength is perceived by the normal eye as yellow). If the light was emitted by a helium atom in a distant star, this happened a long time ago and if the unit length has increased in the meantime by a factor of say 1.5, the wavelength of this characteristic helium line is now 1.5 times larger (and hence perceived as read). Such a "red shift" of spectral lines in distant galaxies had

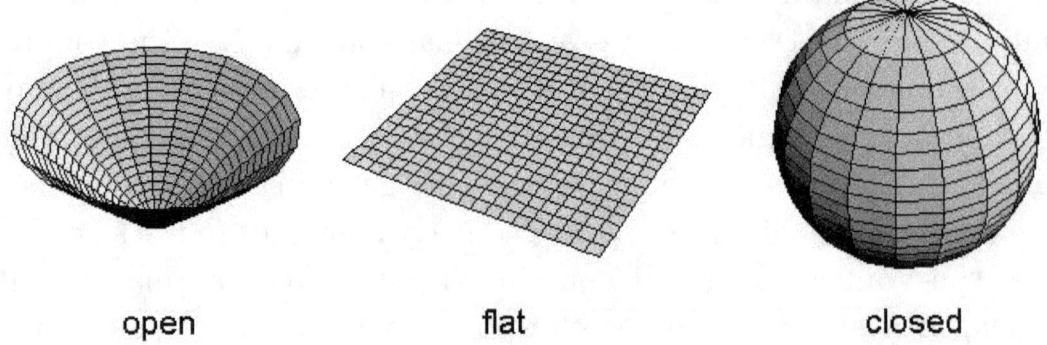

open   flat   closed

Figure 7.12: A two-dimensional open, a flat, and a closed *two*-dimensional space, embedded in three-dimensional Euclidean space. The open space corresponds to $k = -1$, the flat one to $k = 0$ and the closed one to $k = +1$.

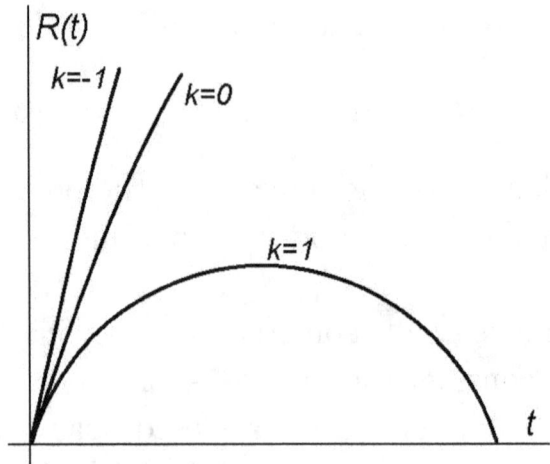

Figure 7.13: Time behaviour of the unit length $R(t)$ for a closed ($k = 1$), asymptotically flat ($k = 0$) and open ($k = -1$) universe.

been first observed by the American astronomer E. P. Hubble in 1929 for galaxies relatively near to our one and has since been observed for galaxies over a vast range of distances. As had been conjectured already by Hubble on the basis of his first findings, the red shift increases with the distance of the object emitting the light, the further away a galaxy, the faster it moves away from us. This is the famous expansion of the universe, on which we shall say a few words later. In Figure 7.13 we show the behaviour ot the unit length R in time for the three cases $k = 0, 0,_1$ Before we proceed further, we have to explain the equivalence of the two seemingly different approaches, namely that of an universe where the distance between the objects increases and that of an increasing unit length. They are indeed equivalent. If we measure the distance in present day units than the distance between the galaxies increases, they move away from us and we obtain the red shift from classical physics as the so called Doppler shift[12]. In general relativity it is more appropriate to express distances in the actual unit length. If this unit length (expressed e.g. in present-day unit lengths) changes with time, distances staying the same if measured in the actual unit lengths, increase if they are expressed in present-day units. The same increase occurs for the value of the optical wavelength of a characteristic spectral line. So we see that we can easily explain the Doppler effect also in the language of general relativity.

In order to measure the relative rate of change of the unit length ($\dot{R}/R$), which enters the Friedman-Lemaître equations, we must measure the change of wavelength of characteristic lines emitted from a certain object, which is easily achieved with high precision and – and this is extremely difficult – we must know the distance of the object emitting the light. The present-day value for the relative rate of change of the unit length, the so called Hubble constant, is approximately 2.23 cm/s per lightyear, that is an object in a distance of 1000 lightyears moves away with a velocity 2.23 km per second. Equivalently, the unit length changes in one million years by about less than one hundredth of a percent.

Given the present-day value of the Hubble constant one can calculate from the Friedmann-Lemaître equations a critical cosmological density which determines the large scale curvature of the universe. If the total density, which includes the cosmological constant, is larger than the critical density, the constant $k = 1$, the curvature is positive and the universe is closed. Here some remarks on the "expanding universe are in order: In the model most favored by observational data, that is the asymptotically flat universe, the universe is infinite in space and therefore cannot increase in size. What indeed is increasing is the unit length, and therefore the distance between objects if measured in units of a fixed time. Only for the case that the unit length would strictly take the value 0 one could speak of an universe

---

[12]The Doppler shift was first detected in acoustics by the Austrian physicist Ch. Doppler in 1842. He demonstrated that a clarinet playing a certain note sounded higher when it was approaching the listener and sounded lower when it was moving away. Listeners which perfect pitch could even determine the speed of the player (seated on a railway wagon).

without extension at that time but jump to infinite size immediately after.

Besides the Hubble "constant" (which is indeed in general a time dependent quantity) the cosmic-microwave-background (CMB) radiation is another extremely valuable source of information for the structure of the universe[13]. The cosmic microwave background is an extremely isotropic electromagnetic radiation with a distribution of wavelengths corresponding to the thermal radiation of a body of ca 3 degrees above absolute zero (-270 degree centigrade). The existence of such a background radiation had been predicted by G. Gamow and coworkers as early as 1948. Gamow was interested in the nuclear synthesis of the chemical elements and assumed for that purpose a primordial fire ball[14] where the nuclei of chemical elements were formed. This fire ball would cool in an expanding universe by the redshift of the electromagnetic radiation and the present-day temperature was estimated to be around 5 degrees above absolute zero. Although it was found later that the heavier elements (such as oxygen and iron, e.g.) are synthesized in the interior of stars, the interest in a possible cosmic microwave background was revived in the 1960s and experiments to detect it were planned[15]. The actual discovery came however from Bell Telephone Laboratories through an experiment dedicated to other purposes, namely those connected with the first radio signal transmission via satellites. Electromagnetic back ground noise to the transmitted signals comes mostly from our galaxy, that is principally from the region of the milky way. But much to their astonishment, A. Penzias and R.W. Wilson found in the early 1960s also a very isotropic long-wavelength radiation corresponding to a temperature of about 3 degrees above absolute zero. By discussions with cosmologists it was soon recognized that this very homogeneous radiation was the back ground radiation predicted by Gamow and searched for by R.H. Dicke and coworkers at Princeton.

From the extreme isotropy of the cosmic microwave background one can conclude that somewhere at an early stage of the universe the expansion (increase of the unit length) was extremely fast, this was the

---

[13]The Belgian cleric and cosmologist G. Lemaître called it "the vanished brilliance of the origin of the worlds"

[14]The name "big bang" was introduced by F. Hoyle, a firm supporter of a steady state theory, in a rather derogatory way

[15]In 1948 the detection of such a low temperature (long wave length) radiation seemed to be impossible

so called inflationary phase. From tiny deviations from homogeneity one can conclude that the universe is asymptotically flat since these inhomogeneities indicate that the sum of angles in a triangle is 180 degrees, as it is in Euclidean geometry. This is very consistent with the inflationary phase mentioned above, since inflation would exactly tune the density to the critical one and hence predicts $k - 0$.

If one follows back the development of the unit length using the Friedmann-Lemaître equations it turns out that about 13 billion years ago the unit length was was tiny compared to the present one. At this time the Higgs mechanism was not yet operative and particles had no mass. The temperature was so high that the quadratic term of the champagne bottle potential (see 7.7) was more than compensated by the kinetic energy. But even after the Higgs mechanism became operative and particles acquired a mass, there were not yet hadrons. The strongly interacting quarks and gluons were in plasma in which they couldn't move freely, but were not yet confined inside hadrons. Hadrons could form only at temperatures corresponding to an average energy of 100 MeV, this temperature was reached approximately a tenth second after the big bang. After about three seconds the universe had cooled off to a temperature of half a MeV and the known particles like electrons, protons, neutrons "froze out". All unstable particles like mesons, strange hyperons and muons had decayed long since.

The ratio of protons to neutrons of about six to one was determined by the chemical equilibrium at the "freezing temperature" of ca 1 MeV. From then on during the first three minutes the light 'primordial' elements were formed. They range from heavy hydrogen (one proton and one neutron) to Lithium (three protons and four neutrons). Besides hydrogen, helium is the by far most abundant element in the universe. Nearly a fourth of all nucleons in the universe are bound in the helium nucleus. The reason for that abundance is the very high binding energy of the nucleons inside its nucleus. Lithium is very scarce, for one lithium nucleus there are more than 2 million hydrogen nuclei. From these numbers on can conclude that for each nucleon in the universe there are 10 billions of photons, but these photons contributed only for a very short fraction of the age of the universe significantly to its (energy) density. This is due to the low energy of the photons. Due to the expansion of the universe they have cooled of from a once dominant part of the universe to the meagre background radiation, a photon gas with a temperature of only 2.7 degrees centigrade above absolute zero. This corresponds to an average photon energy of two thousandth of an electron volt.[16]

The conventional matter, of which we have sure knowledge, that is all the particles occurring in the Standard Model, contributes only to a small fraction to the density of the universe. Its complete density can be determined from different sources, e.g. from the speed of the expansion of the universe and from the rotation of spiral nebulæ. The conclusion from observations made in the last years is very exciting: About 30% of the universe is dark matter, the rest of ca 70% is due to the cosmological constant, sometimes called quite inadequately "dark energy". The dark matter makes itself felt through a retardation, the cosmological constant through an acceleration of the expansion of the universe. Dark matter consists only to 10 % of conventional matter, the rest is unknown to us.

All these numbers are taken from model calculations, but the fact that completely independent methods lead always to compatible numbers gives the model calculations a high degree of reliability.

In relativistic quantum field theory, the basic theory of the Standard Model of Particle Physics, one has no problem to explain the existence of cosmological constant in the universe. The problem is the opposite: the cosmological constant expected in the Standard model is far to large. Quantum corrections to the vacuum energy are unavoidable and they lead to an energy density of the vacuum. It is only detectable through its interacting with the gravitational field. In renormalized perturbation theory there is no way to calculate these quantum contributions. In a field theory without gravity the vacuum-energy density has no influence and therefore it is not possible to fix it at a certain scale by a measured value. If one supposes, however, that quantum field theory gives a realistic description of nature up to a smallest length, then the calculated vacuum energy has to be taken seriously. If one assumes that the theory is valid down to distances corresponding to the Planck mass, that is when quantum effects of gravitation are supposed to become important, then one obtains for the energy density of the vacuum and thus cosmological constant a value of the Planck mass to the fourth power. Expressed in GeV per cubic femtometer, this is $2.6 \cdot 10^{59}$ GeV/fm^3. The observed value for the dark energy density is ca $6.6 \cdot 10^{-45}$ GeV/fm^3, that is a discrepancy

---

[16]Nevertheless the photons outnumber the nucleons (protons and neutrons) by about two billions $(2 \, 10^9)$ to one

of one hundred orders of magnitude, an absolute record in misjudgement.

This already shows that there is no proper match between general relativity and quantum theory, and indeed there exists at the moment no consistent quantization scheme of gravity within the Standard models. Only if we ignore these difficulties and extrapolate the Friedmann-Lemaître equations as they stand to the time 0 we obtain for the unit length the value zero. This (hypothetical) event is called the "Big Bang". It is indeed a remarkable singularity, since at this moment the universe had no extension at all and before it did not exist at all, neither did space and time. In Figure 7.14 we show two curves for the development of the unit length $R(t)$. The black curve is the solution of the classical Friedmann-Lemaître equations, the gray curve is one based on models which try to unify quantum physics with the theory of gravity based on the so called superstring model. In it a minimal length, the so called Planck length exists which however is extremely small even on the scale of elementary particle physics.[17] The minimal length which can be resolved by the huge modern particle accelerators is about 1 million billions times larger than this Planck length. The gray curve in Figure 7.14 based on superstring theory shows no singularity and therefore a "pre-Big-Bang" history of the universe. Though the curve is the result of very speculative model considerations, it is in a sense better than the black curve inside the gray region, since it at least tries to resolve the inconsistencies inherent in the classical extrapolation. Nonetheless the time where quantum corrections played an important role is extremely short, a crude estimate yields the unimaginably short time of 10^{-40} seconds. The time A B B (after Big Bang) after which we are rather certain to know the physical laws and after which we can trace the history rather reliably is about 10^{-9} seconds (a nanosecond). The most distant observed objects are about 10 billion light years away, i. e. the time span which is "directly" accessible to observation started only several billion years A B B.

Since it is one of the primary aims of this book to show the importance of interface building between different branches for the development of science we shall take the opportunity in a very short recapitulation of the cosmological history of the universe to stress es-

---

[17]The Planck length, closely related to the Planck mass, is given by $L_P = \sqrt{\hbar G_N/c^3} \approx 1.6 \, 10^{-35}$m, where $\hbar$ is the Planck constant, G_N Newton's gravitational constant, and c the velocity of light.

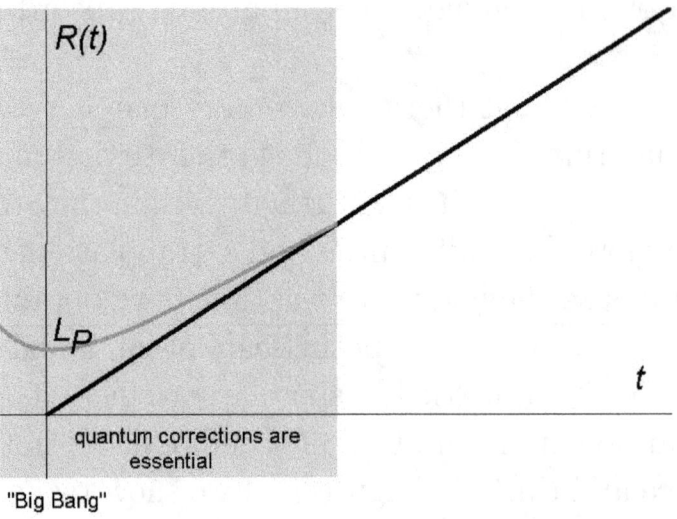

Figure 7.14: Extrapolation of the classical Friedmann-Lemaître equations to $t = 0$ leads to the singular geometry with $R(0) = 0$. Inside the gray region the classical equations are however not reliable and the singularity may never occur. The gray curve is an example of such a "pre-big-bang scenario"

sentially this aspect. There is however a strong caveat. As we have seen above, only about a small percentage of the matter in the universe is conventional matter, that is matter accounted for in the Standard Model of Particle Physics. It is evident that the unknown parts of the universe may have unsuspected effects.

We start the history with the – most probably hypothetical – Big Bang, that is the extrapolated point where the unit length becomes zero if one continues the solutions of the classical Friedmann-Lemaître equations into a region where they are surely no longer valid.

The first period until about 10^{-40} seconds A B B are dominated by physical laws unknown to us. It may be the only time in the whole history of the universe where quantum effects in gravity were important. Therefore this region is the preferred playground of physics far beyond the Standard Model of Particle pPhysics. The explanation of the behaviour in the universe in that early period is for instance one of the principal motivations for superstring theory. In such a theory a big bang does most probably not occur at all and one can construct pre-big-bang scenarios.[18] There is only scarce[19] hope to get

---

[18]it is interesting to note, that the term scenario, which was originally a technical term in the *commedia del arte* indicating a rough sketch of the action became also very popular in physics

[19]we say scarce, since in science one should never say *no*

real constraints on models from this period, it is mostly a question of consistency which motivates theoretical investigations there.

The next period, which extends until a nanosecond (10^{-9} seconds) A B B is a period governed by physics beyond the Standard model. Here important constraints on theory may be obtained from cosmology. This is especially true since there is a candidate for an extension of the Standard model of particle physics, the so called supersymmetric Standard model, which has a good chance to cover most of this period. The results expected to come from the new accelerator LHC at CERN may give some very important hints on the validity of this model in the next couple of years.[20]

The following period, ending about 3 milliseconds A B B, is the period where the physics of the Standard model in its full glory is dominant. It is the era where the Higgs mechanism (see 7.1.2) occurs and quarks become an important part of conventional matter. It can, however not (yet) be explained why there is evidently only matter and no antimatter in the universe. This is one of the big challenges of the two Standard models. The experiments at LHC are expected to give important information concerning this question.

The next three minutes were an era where nuclear physics became important, it was essential for the formation of the primordial elements, especially of the nuclei of the element helium out of neutrons and protons. The relative abundances of the light elements: (normal) hydrogen, heavy hydrogen (deuterium), helium, and lithium yield one of the most important consistency checks of the Standard model of cosmology.

Another important event in the physical history occurred about 800 000 years A B B. In this time the electrons were captured mostly by the atomic nuclei and the stable atoms were formed. At the same time the radiation density (particles without rest mass) and the matter density decoupled. Just before the decoupling time the universe behaved essentially as a proton-electron plasma tightly coupled to a photon gas. The present status of the cosmic microwave background reflects the state of the plasma at this time, since the only change of the cosmic microwave background after this decoupling came from the expansion of the unit scale. The fine structure of the cosmic microwave background is therefore an excellent indicator of the universe when its

---

[20]Since one has found the Higgs boson, but no significant indications for supersymmetric partners of known particles, this hope is diminished

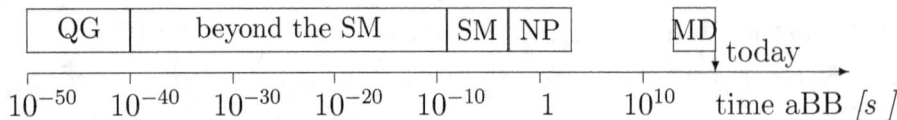

Figure 7.15: Different periods of the universe plotted against time in seconds after the Big Bang (aBB). **QG**: era where quantum gravity was important; **beyond the SM**: era, where physical laws beyond the Standard model of particle physics were important; **SM**: era, where physical laws of the Standard model of particle physics were important; **NP**: era where nuclear reactions were dominant and where primordial elements were synthesized; **MD** : matter dominated era.

state was relatively simple. The information that the universe is most probably asymptotically flat ($k = 0$) stems from a sophisticated analysis of the small inhomogeneities of the (present) cosmic microwave background.

Since matter with rest mass cools down much slower than radiation, the radiation density played from that time on a less and less important role in the total density of the universe.

As mentioned above, the oldest objects we have observed so far emitted their light about 3 billion years A B B

The different eras are shown schematically in Figure 7.15.

For readers interested in numbers we collect some important cosmological parameters.

The present day value for the temperature of the cosmic microwave background is 2.725° Kelvin, that is 2.725 degree centigrade above absolute zero temperature corresponding to about -270 degree centigrade on the Celsius scale.

The present day value of the Hubble constant, the relative rate of change of the unit length, is : $H = 1/10$ billion years.

It is convenient to introduce the normalized densities Ω_x, where the subscript x stands for various kinds of "matter" ($x = m$: slowly moving particles in general with rest mass, $x = b$: nucleons, $x = r$: electromagnetic radiation) and the "vacuum density" Ω_v [21] in such a way, that if $\Omega_{\text{tot}} = \Omega_m + \Omega_r + \Omega_v$ equals one, the universe is asymptotically flat (k=0). Various observations lead to the following values: $\Omega_m \approx 0.24, \Omega_v \approx 0.76$. The contribution of normal matter, that is

---

[21]$\Omega_x = \frac{8\pi G_N \rho_x}{3H^2}$, $\Omega_v = \frac{\Lambda}{3H^2}$ where ρ_x is the respective density repectively and Λ is the cosmological constant

matter occurring in the Standard Model and what would have been called matter in general some twenty years ago, is only about 20% of the total matter density Ω_m. The present day value of radiation density is negligibly small, the contribution of neutrinos is expected to be small, but its precise value is still open.

Chapter 8

EPILOGUE

I

Instead of asking the question what is science, one might ask: How is it distinguishable from non-science? It was Sir Karl Popper (1934/1959) who proposed the principle of falsifiability as a criterion of demarcation to answer this question. The principle of falsifiability requires that, unlike non-science, a science should be able to produce falsifiable theories. Otherwise it is not science. What is most important in this context is not the dynamics of theory development in science but the rationality of a candidate theory T as being determined by its logical relationship with its potential falsifiers. The falsifiability of T is, according to this proposal, believed to be demonstrable by means of methodological criteria being employed strictly within the framework of *modus tollens* of classical logic, using statements of appropriate logical form as potential falsifiers against T.

During the second half of the twentieth century, the criterion proposed by Popper has been interrogated and widely debated by the philosophers of science (Pandit 1991). Today, one may or may not accept the falsifiability criterion as a simple logical approach to solving the demarcation problem. At best the criterion of falsifiability articulates the methodological perceptions of epistemologists, such as Popper, who were responding to the scientific developments of the first three decades of the 20th century within physics. In today's scenario of scientific developments, one might instead argue as follows: "it is much more akin to the implicit perception the physicists themselves have of their own science,"[1] that without an active frontier full of new theoretical scenarii, theory-experiment interfaces and new probing re-

---

[1] We borrow here the words of Bernard d'Espagnat (personal communication, 25th July 2006).

search questions on the horizon, a science would not be distinguishable from non-science. The central question which we have been concerned with in this book is this:

> What kind of scientific reasoning is it within physics which triggers the dynamics of problem and theory development at its frontiers, keeping them open to scientific change?

We have answered this question by drawing attention to the scientific reasoning within physics which appeals to the dynamic core-context building resolving power of physical theory. Thus we have argued that, at any time in their development, physics and other natural sciences are characterized by two things: First, each individual field of science will be characterized by dynamic core-context building from *within* it. Secondly, it will be characterized by an active frontier where new research questions can be formulated and explored, given the dynamic core-context. The two, the dynamic core-context and the frontier of research, are dynamically correlated as follows. In order that there are active frontiers in each field of study within physics where received wisdom can be questioned and research questions can be formulated, or explored, and where new experimental facilities can be constructed, the existence of effective dynamic core-contexts is necessary. The former cannot come into being without the latter. That is to say, there can be no frontiers full of research questions without the dynamic core-contexts comprising of successful relevant theories, theory-experiment interfaces, laws and experimental results, which jointly build up the *context* of scientific reasoning, making it meaningful to speculate about the future developments or scenarii. Yet, it is important to note that their mere presence or existence in a field of study may not be enough, if physics is to make progress at its frontiers. In order that there is real advancement of problem and theory development at its active frontiers, instead of mere theory testing, the relevant core-contexts must be effective. This raises the question what is an effective core-context of problem and theory development.

The argument of this book draws attention to the scientific reasoning in physics whereby a dynamic core-context becomes effective. As a complex structure of successful theory and theory-experiment interface, laws and theories and their framework, an effective core-context serves as an essential input to theory development at the active frontiers of physics where challenging new research questions are waiting to be formulated or explored. At present, the Standard Model of particle physics and the Standard Model of cosmology, in their development over the past several decades, are the best examples of an effective core-context in science. The following statement by the Nobel Prize winner M. Veltmann can be interpreted as expressing his excitement on the dynamics of a new core-context of development in his seminal talk on "Gauge Field Theories" in 1973 when the full core of the Standard Model of particle physics was clearly emerging:

> "It is not bon ton in a scientific expose to show excitement, and I will accordingly try to avoid this. Yet it is at this moment very difficult to do so, because it seems more and more that we are on the right track".

Without the effective core-context playing its dynamic role, it would not even be possible to pose new research questions, nor to invest on the scale of the LHC-type new experimental set-up of elaborate search-and-discovery procedures of finding experimentally testable answers to these questions. According to the model of theory development presented in this book, with respect to every newly anticipated advance of theory development in physics, there must be an essential input in the form of an effective core-context, no matter what other core-contexts are present in the neighbourhood at the same time. Thus, what makes the core-context effective are its powerful resources which shape the active frontiers where one is searching for a new theory, which could address those research questions which are waiting to be answered because these cannot be answered within the present core-context. In a nutshell, we have made an attempt in this book to articulate the nature of scientific reasoning at the frontiers of problem and theory development, which is based on, or which appeals to, the dynamic core-context building resolving power of physical theory.

II

At the core of the methodological proposals made by us in this study lies the following view:

> That theories and problems in science, no less than novel experiments including thought experiments, are the developmental structures of knowledge.

As a methodology of scientific progress, this view has, as we shall see in a moment, far-reaching consequences that go even beyond the natural sciences, penetrating other sciences. In particular, our dynamic approach to theories and problems in the sciences, viewing these as the developmental structures of knowledge or as valuable epistemic resources for basic and applied research, is originally traceable to Pandit (1982, 1989, 1991, 2002a, 2002b). Notice that this approach entails the revolutionalry view that knowledge is a highly context-dependent resource rather than context-independent homogeneous absolute "truth". Notice that the conception of knowledge as a context-dependent epistemic resource challenges most of the fundamental assumptions of the most dominant schools of epistemology and epistemic communities that are committed to different kinds of philosophies. On the creative side, at the same time, it promises new developments that can shape innovative approaches to interdisciplinary research that have become imperative for disciplinary interface-building for a sustainable future for humanity. Neither the natural sciences nor the social sciences can miss the important implications this fundamental shift carries for the public understanding of basic scientific research. These sciences cannot miss its implications for the innovative and strategic knowledge resources which they can individually or jointly produce. Knowledge is a valuable epistemic resource which can be used to produce other types of valuable resource. In order that significant improvements and changes can be brought about elsewhere, in the world, in society, in technology, in human condition, in sustainable development, knowledge must itself be revisable. As a product of basic research, knowledge (or new technologies) can be fed into the interdisciplinary research to produce the sustainable and efficient solutions to tackle the greatest challenges confronting the humanity at present. Thus, our study can be judged in terms of its frontier crossing, far-reaching consequences for the information society of the 21st century. As we have just hinted at, it has far-reaching consequences, e.g., for fostering strategic interdisciplinary research at the newer and most challenging frontiers that have emerged in recent decades, or that are still emerging in the horizon, pushing humanity for more action for building interfaces between scientific disciplines caught in their separate past histories.

III

We have treated in this book to great detail how philosophy should look at physics and science in general, in order to establish a consistent philosophy of science. One can also reverse the question: Has science to consider philosophy? We have mentioned in sect. 2.2.1 the harsh critique of S. Weinberg about the "Unreasonable Ineffectiveness of Philosophy" for a scientist, but also quoted the famous statement of H. Helmholtz "to investigate the sources of our knowledge and the degree of its justification is a duty, which will ever be that of philosophy and no century can avoid that unpunishedly".[2] Einstein, certainly a philosophical mind, put, however, the acceptance of the results of "this duty" into perspective with the principal task of a scientist and wrote in his "Reply to criticism" (Einstein 1951, p. 684):

> He ⟨The scientist⟩ accepts gratefully the epistemological analysis; but the external conditions, which are set for him by the facts of experience, do not permit him to let himself too much restricted in the construction of his conceptual world by the adherence to an epistemological system. He therefore must appear to the systematic epistemologist as a type of unscrupulous opportunist.

Nevertheless, we think that the theory of signs and symbols initiated by such eminent scientists as Helmholtz, Hertz and Poincaré (see Dosch 1997, Dosch et al. 2005) and brought into an impressive system by the philosopher E. Cassirer (1923, vol. 3), offers a frame, which on one hand does not restrict too much the conceptual world, but on the other hand does prevent the scientist from getting lost into too wild an "epistemological opportunism".[3]

This is illustrated by the problem of "the existence of quarks": About the question "Do quarks exist?" one can have many fruitless discussions, which have nothing to do with quarks, but only depend on the interpretation of "exist". [4] On the other hand, no physicist could object to the statement that quark fields are extremely useful symbols in the framework of particle physics, see sect. 7.2. In the light of the Maldacena conjecture (sect. 7.2.5) it is thinkable that this symbol can

---

[2]H. Helmholtz (1884, Bd I, p.368)
[3]see e.g. Dosch, 1990, Dosch et al. 2009.
[4]see e.g. Dosch et al. 2011

be given up in certain aspects of particle physics, aspects which might be more appropriately described in a 5-dimensional theory.

IV

In conclusion, we would like to further clarify why it is inappropriate to view fundamental processes of scientific change taking place within the natural sciences such as physics by means of the highly systematically ambiguous concept of a paradigm, which has been popularized by Thomas Kuhn (1962) and Kuhn scholars. Around the middle of the twentieth century, this concept was brought into the philosophical discourse on language, and on forms of life/activity/rule-following behaviour, as part of culture, that are inseparably associated with *it*, by later Wittgenstein in his pothumously published work entitled *Philosophische Untersuchungen/Philosophical Investigations* (PU 1953). Although this development went directly against his own earlier work (TLP), it clearly signalled a paradigm shift of sorts within the twentieth century anlytical philosophy. With its far-reaching implications, this development should have transformed humanities and the social sciences, and all discourse on them, beyond recognition. On the contrary, one witnessed the historians of science resorting to a kind of Wittgensteinian strategy, seeking to transform the historical and public understanding of scientific change, while reconstructing the scenarii of scientific revolutions within the natural sciences like physics.

However, with his new approach to language, bringing in the concept of a paradigm with multiple roles in his PU (1953), Wittgenstein had aimed at teaching us how mistaken it is to articulate concepts, e.g., language, game, knowledge, truth, meaning and the like, within the boundaries of an essentialistic/absolutistic abstract definition, particularly in the kind of situation where one has the tendency to ask the following abstract questions: What is language? What is knowledge? What is meaning? What is truth? It must have been his intention, then, to suggest how much the highly systematically ambiguous concept of a paradigm was suited to the philosophical tasks of articulating the fundamental yet difficult and open concepts or conceptual frameworks.

Thus, the historians of science, notably Thomas Kuhn, hastened to transfer the concept of a paradigm to their projects of articulation of the history of scientific revolutions and scientific change. We must not forget that while it might be quite useful to do so with respect to the social sciences or humanities, it is a serious mistake to employ such a highly systematically ambiguous concept in connection with the natural sciences, more so in the context of the twentieth century developments in physics and cosmology. Our reason for saying this is both simple and complex. In the original context where Wittgenstein employed it, the concept of a paradigm is entangled with considerable metaphysical, not just ideological/sociological, baggage, which may also vary with variable contexts of its use. No surprise, if we find so much literature devoted to science studies, authored by Kuhn scholars, which focuses on the many meanings of the term 'paradigm', or on in-commensurability between paradigms, or on meta-incommensurability issues. Kuhn himself responded to much criticism of his own approach to articulating 'scientific revolutions by recognizing several senses in which one could use the term 'paradigm', and himself used it in at least two senses, which we have considered in chapters 2 and 6.

Within the science studies, one cannot transfer a highly systematicaly ambiguous concept, such as the concept of a paradigm, to any of the contexts of discourse on the natural sciences, howsoever inter-disciplinary in nature it may be, without bringing in its metaphysical and ideological/sociological baggage. In the case of discourse on the social sciences or humanities, this should not pose any formidable dif-ficulty nor come in the way of rigorous articulation of the fundamental processes of social scientific change.

Put in a nutshell, while in the context of the social sciences it may make sense to show scientific change as a kind of paradigm change, in the context of the natural sciences, it makes no sense at all. This may explain, independently of the arguments developed in this book, why it is of fundamental methodolognal importance, *in their context*, to pay attention to the scientific reasoning from the core-context build-ing resolving power of physical theory, which we have attempted to articulate in terms of core-contextual methodological structuralism.

Bibliography

[1] Anderson, C. D. (1932), *Science* 76 (1932) 238

[2] Anderson, J. D. et al (1998): *Phys. Rev. Lett.* 81 (1998)

[3] Arnold, V. I. (1978): Mathematical Methods of Classical Mechanics, Springer Verlag: New York/Heidelberg/Berlin.

[4] Bando, M., Kawabe, R. and Nakanishi, N. (eds.) (1985) "Proceedings of the Kyoto International Symposium: The Jubilee of the Meson Theory (Kyoto, August 15-17, 1985)", in Progress of Theoretical Physics - Supplement, N0. 85, 1985

[5] Beaty, J., Cartwright, N. et al., "Introduction to Volume 2", in: The Probabilistic Revolution Vol. 2, ed. Krueger, Lorenz et al. (Cambridge, Mass./London, 1987), pp. 1-4.

[6] Bell, J. S. (1988): *Speakable and Unspeakable in Quantum Mechanics*. Cambridge: Cambridge University Press. (Collected Papers on Quantum Philosophy, Ist ed. 1987)

[7] Beller, M. (1999), *Quantum Dialogue*, University of Chicago Press: Chicago & London.

[8] Blackwell, R. J. (1984) "Review of G. L. Pandit, The Structure and Growth of Scientific Knowledge", D. Reidel: Dordrecht/Boston/London, 1982, in The Review of Metaphysics 38, 1(149), 1984, pp.673-674.

[9] Bohm, D. and Peat, F. D.(1987): *Science, Order and Creativity*. London: Routledge.

[10] Bohr, N., "Can Quantum Mechanical Description of Physical Reality be Considered Complete?" Physical Review **48** (1935), pp. 696-702.

[11] Boltzmann, L, Populare Schriften (Leipzig, 1905). Cohen, I. B., "Preface" to Newton, I., Opticks (New York, 1952). An unabridged and unaltered republication of the 4th edition (London, 1730), IX-LVIII.

[12] Bourbaki, N. (1984), Elemente d'histoire des mathematiques, Paris: Massan, 1984.

[13] Calvin, H. (2001) Fallibilism, Democracy and the Market: The Meta-Theoretical Foundations of Popper's Political Philosophy. Lanham/New York/Oxford: University Press of America, Inc.

[14] Carnap, R.(1934) Logische Syntax der Sprache. Schriften zur Wissenschaftlichen Weltauffasung, herg. von Philipp Frank und Moritz Schlick, Bd. 8. Wien: Verlag von Julius Springer, xi+274 pp.

[15] Carnap, R. (1937), The Logical Syntax of Language (orig. German ed. 1934 Vienna). London/New York: Routledge & Kegan Paul.

[16] Carnap, R. (1936/1937) "Testability and Meaning", Philosophy of Science (Baltimore) Vol. 3 (no. 4, October 1936) pp. 419-471; Vol. 4, (no. 1, January 1937) pp. 1-40. Reprinted as:

[17] Carnap, R. Testability and Meaning, New Haven, Connecticut 1954.

[18] Carnap, R. 1950. "Empiricism, Semantics, and Ontology", Revue Internationale de Philosophie 4: 20-40. Rept. in Carnap, R. 1956, Meaning and Necessity. Chicago: University of Chicago Press.

[19] Carnap, R. (1938), "Logical Foundations of the Unity of Science" in *Encyclopedia of Unified Science* (Foundations of the Unity of Science Vol 1 (1), 1938: 42-62, Chicago: University of Chicago Press

[20] Carnap, R. (1950), *Logical Foundations of Probability*, Chicago: University of Chicago Press (2nd ed., 1962).

[21] Carnap, R. (1956a), *Meaning and Necessity: A Study in Semantics and Modal Logic* (1st ed. 1947). Chicago: University of Chicago Press.

[22] Carnap, R. (1956b), "Methodological Character of Theoretical Concepts", in Herbert Feigl, et al. (eds.) Minnesota Studies in the Philosophy of Science Vol. I, pp. 38-76. Minneapolis: University of Minnesota Press.

[23] Cassirer, E (1923), *Die philosophie der symbolischen Formen*, Berlin: B. Cassirer.

[24] Cassirer, E. (1937) *Determinismus und Indeterminismus in der modernen Physik*, as cited in

[25] Cassirer, E. (1946/1953) *Language and Myth*, S. K. Langer (trans.). Dover Publications: New York.

[26] Chandrasekhar, S. (1974), "Development of General Relativity", Nature 252 (15-17)

[27] Chew , F. C. (1961), *S-Matrix Theory of Strong Interactions*, Benjamin: New York.

[28] Cohen, R. S. (1968), "Ernst Mach: Physics, Perception and the Philosophy of Science", in *Synthese* 18 (1968), pp. 132-170.

[29] Cohen, I. B., "Preface" to Newton, I., Opticks (New York, 1952). An unabridged and unaltered republication of the 4th edition (London, 1730), IX-LVIII.

[30] Cohen, I. B., Revolution in Science (Cambridge, Mass., 1985).

[31] Coleman, S. and Mandula, J. (1967) "All Possible Symmetries of the S Matrix", in *Phys. Rev.* 159: 1251.

[32] Conversi, M., Pancini, E. and Piccioni, O. (1947) , *Phys. Rev.* 71 (1947) 209

[33] Damour, T. and Taylor, J. H. (1992), "Strong-field tests of relativistic gravity and binary pulsars" Phys. Rev. D, 45: 1840-1868, 1992.

[34] Damour, T. (2006a) "Experimental Tests of Gravitational Theory", Physics Letters B: Review of Particle Physics 592: 186-190. Issues 1-4

[35] Damour, T. (2006b),*Once upon Einstein*, A. K. Peters, Wellesley 2006

[36] Delacre, G. (1984). "Review of G. L. Pandit, The Structure and Growth of Scientific Knowledge", in Dialogos ANO 19 (Numero 44), 1984, pp. 189-194. (in Spanish).

[37] D'Espagnat, B. (2006), On Physics and Philosophy, Princeton University Press.

[38] Devitt, M. (2001), "Incommensurability and the Priority of Metaphysics" in Hoyningen-Huene, P., et al. (eds.): 2001, pp. 143-157.

[39] Dilthey, W. (1962), *Gesammelte Schriften*, Stuttgart-Goettingen, Vol. 1: Einleitung in die Geisteswissenschaften, 5th ed.

[40] Dirac, P. A. M., (1930) , Proc. Roy. Soc. A126 (1930) 360, A 133 (1931) 60

[41] Dirac, P. A. M.(1978), "The Development of Quantum mechanics", in H. Hora and J. R. Shepanski (eds.) *Directions in Physics* (lectures delivered by P. A. M. Dirac August/September1975). New York: John Wiley & Sons, 1978.

[42] Dirac, P. A. M. (1983), "The origin of quantum field theory" (pp. 39-55) in Laurie. M. Brown and Lillian Hoddeson (eds.) *The Birth of Particle Physics*. Cambridge: Cambridge University Press, 1983.

[43] Dosch, Hans Günter (1990), "Cassirer's Erkenntnistheorie, Kommentar eines Physikers", in *Determinismus - Indeterminismus: Philosophischc Aspcktc physikalischer Theoriebildung*, Herausgegeben von Wolfgang Marx . Vittorio Klostermann: Frankfurt am Main, S. 111-135.

[44] Dosch, H.G. (1997) "The Concept of Sign and Symbol in the Work of Herman Helmholtz and Heinrich Hertz", in *Cassirer 1945 - 1995, Sciences et Culture*, N. Janz ed., Études de Lettres, Lausanne 1997

[45] Dosch, H. G., Müller, V.F. and Sieroka, N. (2005b), *Quantum Field Theory in a Semiotic Perspective*. Springer: Berlin/Heidelberg/New York, 2005.

[46] Dosch, H. G., Müller, V.F. and Sieroka, N. (2009), *Symbolic Constructions in Quantum Field Theory*, in M.Bitbol et al. ed., *Constituting Objectivity – Transcendental Perspectives on Modern Physics*, Springer,Berlin, Heidelberg, New York 2009

307

[47] Dosch, H. G. (2007), "The Standard Model of Particle Physics", in Seiler, E. et al. (eds.) *Approaches to Fundamental Physics*, Lecture Notes in Physics 721, Springer, pp. 21-50.

[48] Dosch, H. G. (2008), *Beyond the Nanoworld: Quarks, Leptons, and Gauge Bosons*, AK Peters, Ltd: Wellesley, Massachusetts.

[49] Dosch, H. G., Müller, V.F., (2011)"The Facets of Quantum Field Theory", Eur. Phys. Journal H, 35 2011, 331-376

[50] Dosch, H. G. (2010), " Formale Sprache und Umgangssprache in den Naturwissenschaften", in *Wissenschaft und Gesellschaft, Begegnung von Wissenschaft und Gesellschaft in Sprache*, herausg. P. Kirchhof, ed., Winter, Heidelberg, 2010, p. 45-55

[51] Duhem, P. (1906). La theorie physique: son object et sa structure. Paris: Chevalier & Riviere, Editeurs. PP. 1-450 (Bibliotheque de philosophie Experimentale II), 2me ed. 1914. English translation: *The Aim and Structure of Physical theory*, P. P. Wiener (transl.), Princeton University Press, Princeton, 1954.

[52] Einstein, A. (1916), *Phys. Z.* XVII (1916), p. 101.

[53] Einstein, A. (1919), *Über die spezielle und die allgemeine Relativitätstheorie.* Vierte Auflage: Fried. Vieweg & Sohn, Braunschweig, Germany, S. 52. (First German ed. 1916).

[54] Einstein, A. (1960), *Theory of Relativity, Special and General*, p. 77.

[55] Einstein, A. (1934a), *Mein Weltbild*, Herausgegeben von Carl Seelig. Ullstein Buecher.

[56] Einstein, A. (1934b), *Essays in Science* (Philosophical Library. 1934).

[57] Einstein, A. (1951) "Reply to Criticism", p. 684 in Schilpp, P. A. ed. (1951) *Albert Einstein,Philosopher-Scientist*, Tudor, New York 1951

[58] Einstein, A., Podolsky, B. and Rosen, N. (1935), "Can Quantum-Mechanical Description of Physical Reality be Considered Complete?", *Physical Review* 47 (777-780).

[59] Einstein, A.(1954) Ideas and Opinions, (New York, 1954).

[60] Feigl, H. et al (eds.): 1956, *The Foundations of Science and the Concepts of Psychology and Psychoanalysis*, Minnesota Studies in the Philosophy of Science, Vol. 1. Minneapolis: University of Minnesota Press.

[61] Feigl, H. and Maxwell, G. (eds.) 1961, *Current Issues in the Philosophy of Science*, New York: Holt, Rinehart & Winston.

[62] Fermi, E., *Z. Phys.* 88 (1934) 61

[63] Feyerabend, P. K. (1957) "On the quantum-theory of measurement", in Observation and Interpretation (A symposium of philosophers and physicists) ed. By S. Koerner in collaboration with M. H. L. Pryce.London: Butterworths Scientific Publications, pp. 121-130.

[64] Feyerabend, P.K. 1962, "Explanation, Reduction and Empiricism" in H. Feigl and G. Maxwell, eds., Minnesota Studies in the Philosophy of Science, Volume 3, Scientific Explanation, Space and Time, pp. 28-97, Minneapolis: University of Minnesota Press.

[65] Feyerabend, P. K. (1965) "Problems of Empiricism", in Beyond the Edge of Certainty ed. by R. G. Colodny. Prentice-Hall, EnglewoodCliffs, N.J., pp. 145-218.

[66] Feyerabend, P. K. (1975) Against Method: Outline of an Anarchistic Theory of Knowledge. Virso/New Left Books: London/New York. Second ed. 1978.

[67] Feyerabend, P. K. (1978) Science in a Free Society. NLB.

[68] Feyerabend, P. K. (1981a) Philosophical Papers. Volumes I/II, Cambridge University Press.

[69] Feyerabend, P.K. (1981b) Realism, Rationalism and Scientific Method: Philosophical Papers, Volume 1, Cambridge: Cambridge University Press.

[70] Feyerabend, P.K. (1981c) "Explanation, Reduction and Empiricism", in Feyerabend (1981a), pp. 44-96. Reprint of Feyerabend 1962

[71] Feyerabend, P. K.(1981d) "Problems of Empiricism", Philosophical Papers Vol. 2 (London/N ew York, 1981).

[72] Feyerabend, P. K. (1987) Farewell to Reason. Verso: London/New York.

[73] Feynman, R. P. (1985), *"Surely You're Joking, Mr. Feynman"*, New York/London, p. 232.

[74] Feynman, R. P.(1992), *The Character of Physical Law* BBC, 1965 (Penguin Books, 1992)

[75] Fock (1926) "Über die invariante Form der Wellen und Bewegungsgleichung für einen geladenen Massenpunkt, *Z. Phys.* 39

[76] Franklin, Allan (1999), *Can That Be Right? : Essays on Experiment, Evidence and Science* (Boston Studies in the Philosophy of Science 199). Dordrecht/Boston/London: Kluwer Academic Publishers.

[77] Fraser, G. (2000), *Antimatter: The Ultimate Mirror*. Cambridge: Cambridge University Press

[78] Fredenhagen, K. et al.: 2007, "Quantum Field Theory: Where We Are", in Erhard Seiler et al. (eds.) *Approaches to Fundamental Physics*, Lecture Notes in Physics 721, Springer, pp. 61-87.

[79] Fritzsch, H., Gell-Mann, M. and Leutwyler, H. (1973) "Advantages of the Color Octet Gluon Picture", Phys. Lett. 47B

[80] Fuller, Steve (2000) *Thomas Kuhn: A Philosophical History for Our Times*, University of Chicago Press: Chicago/London.

[81] Galilei, G. 1632. Dialogue Concerning the Chief World Systems. Florence.

[82] Georgi, H. and Glashow, S. (1974) Phys. Rev. Lett. 32 (1974), 438

[83] Giulini, D. (2007), "Remarks on the Notions of General Covariance and Background Independence" in Stamatescu, Ion-Olimpiu and Seiler, E. (eds.) 2007, pp. 105-120

[84] Glashow, S. (1961) *Nucl. Phys.* 22 (1961), 579

[85] Green, Brian, (2000), *The Elegant Universe: Superstrings, Hidden Dimensions and the Quest for the Ultimate Theory.* Paperback /Vintage

[86] Götberg, *Zur modernen Physik*, Darmstadt: Wissenschaftliche Buchgesellschaft 1980, p. 162.

[87] Gross, D. J. (2005) "The Discovery of Asymtotic Freedom and the Emergence of QCD", *International Journal of Modern Physics* A, Vol. 20, No. 25 (2005) 5717-5740).

[88] Haag, R., Lopuszanski, J. and Sohnius, M.,(1975), *Nucl. Phys.*, B88: 257

[89] Hacking, I. (1993), 'Working in a New World: The Taxonomic Solution', in P. Horwich (ed.), World Changes, The MIT Press, Cambridge, pp.275-310.

[90] Hacking, I.(1983), Representing and intervening (London/New York, 1983).

[91] Hacking, I.(1984), "Experimentation and Scientific Realism", in Scientific Realism, ed. Leplin, J. (London, 1984), pp. 154-172.

[92] Hanson, N. R. (1961), "Is There a Logic of Discovery?" in Feigl, H. and Maxwell, G. (eds.) 1961, *Current Issues in the Philosophy of Science*, New York: Holt, Rinehart & Winston.

[93] Hanson, N. R. (1965), *Patterns of Discovery.* Cambridge University Press, first edition 1958.

[94] Heisenberg, W. (1948) "Der Begriff abgeschlossene Theorie' in der modernen Naturwissenschaft" in *Dialectica* 2, 1948 (331-336).

[95] Heisenberg, W. (1969), *Der Teil und das Ganze: Gespräche im Umkreis der Atomphysik.* R. Piper & Co. Verlag: München. PP. 85-101.

[96] Heisenberg, W. (1979), *Quantentheorie und Philosophie: Vorlesungen und Aufsätze.* Herausgegeben von Jürgen Busche. Philipp Reclam jun. Stuttgart. S. 22-41.

[97] Heisenberg, W. (1983), *Encounters with Einstein*, Princeton University Press, 1983, p. 29

[98] Heisenberg, W. (1984), *Gesammelte Werke*. Piper: Mnchen/Zürich. Band I, 1929-1955, S.335-340. Herausgegeben von Walter Blum, Hans-Peter Dürr und Helmut Rechenberg.

[99] Heisenberg, W. (1985), Gesammelte Werke III, 1969-1976, Herausgegeben von Walter Blum, Hans-Peter Dürr und Helmut Rechenberg. Piper: München/Zürich 1985, S. 459-460.

[100] Heisenberg, W.(1958), Physics and Philosophy (London, 1958).

[101] Helmholtz, H. (1892), *Ansprachen und Reden ... zu Ehren von Hermann von Helmholtz*, Hirschwald'sche Buchhandlung: Berlin 1892.

[102] Helmholtz, H. (1847) "Über die Erhaltung der Kraft", Berlin 1847.

[103] Helmholtz, H. (1855) "Über das Sehen des Menschen", Vortrag 1855, in *Vorträge und Reden*, Vieweg: Braunschweig 1884, Bd I, p.368.

[104] Helmholtz,H. (1862), "Über das Verhältnis der Naturwissenschaften zu Gesamtheit der Wissenschaften: Akademische Festrede 1862", in Vorträge und Reden, Vieweg: Braunschweig.

[105] Hempel, C. G. 1950. "Problems and Changes in the Empiricist Criterion of meaning", Revue Internationale de Philosophie 4: 41-63.

[106] Hempel, C. G. 1965, "Empiricist Criteria of Cognitive Significance", in Hempel, C. G. 1965. Aspects of scientific Explanation. New York: Free Press.

[107] Hempel, Carl G. (1965), Aspects of Scientific Explanation, The Free Press: New York.

[108] Hertz, H. (1894), *Prinzipien der Mechanik in neuen Zusammenhang dargestellt*, Leipzig.

[109] Hertz, H. (1956), *The Principles of Mechanics Presented in a New Form*, authorized English translation by D. E. Jones and T. E. Valley, 1879, Dover reprint.

[110] Hertz, H. (1962), *Electric Waves, Being Researches on the Propagation of Electric Action with Finite Velocity through Space*, authorized English transl. by D. E. Jones, Dover Reprint: New York.

[111] Hewish, A., Bell, S. J., Pilkington, J. D. H., Scott, P. F. and Collins, R. A. (1968), "Observation of a rapidly pulsating radio source", *Nature*, 217: 709-713, 1968.

[112] Hoyningen-Huene, P., Oberheim, E., Andersen, H.: (1996), "On Incommensurability". Studies in the History and Philosophy of Science 27: 131-141.

[113] Hoyningen-Huene, P. and Sankey, H. (eds.): 2001, *Incommensurability and related Matters*, Boston Studies in the Philosophy of Science Vol. 216, Dordrecht/Boston/London: Kluwer Academic Publishers.

[114] Hubble, E. (1929), "A Relationship between Distance and Radial Velocity among Extra-Galactic Nebulae," *Proceedings of the National Academy of Science*, p. 168.

[115] Hulse, R. A. and Taylor, Jr., J. H. (1975) "Discovery of a Pulsar in a Binary System", *Astrophys. J.*, 195: L51-L53.

[116] Hulse, R. A. (1993), "The Discovery of the Binary Pulsar", Nobel Lecture, Dec. 8, 1993, Physics 1993 (48-69).

[117] Jones, Caroline A.: Spring 2000, "The Modernist Paradigm: The Art World and Thomas Kuhn", Critical Inquiry, 26: 488-528.

[118] Jones, Richard A. L.: 2006, Review of Ted Sargent (2006) The Dance of Molecules: How Nanotechnology is Changing our Lives, Thunder's Mouth Press, 304 pp., in Nature, 440/20 April 2006, p. 995.

[119] Joos, E., 2000, in Decoherence: Theoretical, Experimental, and Conceptual Problems, edited by P. Blanchard, D. Giulini, E. Joos, C. Kiefer, and I.-O. Stamatescu (Springer, Berlin), pp. 1-17.

[120] Kane, G. & Pierce, A., eds. (2008), Perspectives on LHC Physics, World Scientific Publishing Co.

[121] Kane, G. (2008), "The LHC - A 'Why' Machine and a Supersymmetry Factory", in Kane, G. & Pierce, A. (2008), pp. 1-11

[122] Kaluza, T. (1918) Sitzungsberger Preuß. Akad. d. Wissensch. 465, *Math. Z. 2.*

[123] Kaluza, T. (1921) *Sitzungsberger Preuß. Akad. D. Wiss.*

[124] Kant, I. (1787), 2. Auflage, *Kritik der reinen Vernunft*, (1. Auflage 1781).

[125] Kemmer, N., *Proc. Cambr. Phil. Soc.* 34 (1938) 354

[126] Klein, O.(1926) , *Z. Phys.* 37.

[127] Klein, O. (1938), *On the Theory of charged fields*, International Institute of intellectual co-operation, Warsaw.

[128] Klein, O. and Jordan, P. (1927), Z. Physik 45 (1927), 751

[129] Kolb, E. W. and Turner, M. S. (1990), *The Early Universe*, Addison-Wesley Publishing Company.

[130] Kretschmann, E. (1917) *Ann. Phys.* 53 (1917), 575

[131] Kuhn, T. S. (1959), "Energy Conservation as an Example of Simultaneous Discovery", in *Critical Problems in the History of Science*, Madison 1959. The three other persons involved are Mayer, Joule and Colding.

[132] Kuhn, T. (1962), *The Structure of Scientific Revolutions*, The University of Chicago Press, Chicago

[133] Kuhn, T. (1970a), "Reflections on My Critics" in I. Lakatos and A. Musgrave, eds.(1970), London/New York, pp. 231-278

[134] Kuhn, T. (1970b), *The Structure of Scientific Revolutions*, The University of Chicago Press, Chicago, 2nd enlarged edition.

[135] Kuhn, T. (1971), "Notes on Lakatos", in Buck, R. C. and Cohen, R. S. (eds.): 1971, *PSA 1970 In Memory of Rudolf Carnap*, D. Reidel/Springer: Dordrecht/Boston/London. PP. 137-146.

[136] Kuhn, T. (1974), 'Second Thoughts on Paradigms', pp.459-482: in Suppe, F. (ed.) 1974, *The Structure of Scientific Theories*, Urbana: University of Illinois Press.

[137] Kuhn, T. (1976), "Theory-Change as Structure-Change: Comments on the Sneed Formalism" *Erkenntnis* 10: 179-199.

[138] Kuhn, T. (1977), *The Essential Tension*, The University of Chicago Press, Chicago.

[139] Kuhn, T. (1981), 'What Are Scientific Revolutions?' Occasional Paper # 18, Center for Cognitive Science, The MIT Press, Cambridge. Rept. in L. Krueger, L. Daston and M. Heidelberger (eds.) *The Probabilistic Revolution*, Vol. I, The MIT Press, Cambridge, pp. 7-22.

[140] Kuhn, T. (1983), 'Commensurability, Comparability, and Communicability', in P. Asquith and T. Nickles (eds.), *PSA 1982*, Vol II, Philosophy of Science Association, East Lansing, pp. 669-688.

[141] Kuhn, T. (1989a), 'Possible Worlds in History of Science', in S. Allen (ed.), *Possible Worlds in Humanities, Arts and Sciences*, Walter de Gruyter, New York, pp. 9-32.

[142] Kuhn, T. (1989b), 'Response to Commentators', in S. Allen (ed.) 1989, pp. 49-51.

[143] Kuhn, T. (1990a), 'Dubbing and Redubbing: The Vulnerability of Rigid Designation', in C. Savage (ed.), *Scientific Theory*, University of Minnesota Press, Minneapolis, pp.298-301.

[144] Kuhn, T. (1990b), An Historian's Theory of Meaning', Talk to Cognitive Science Colloquium, UCLA (unpublished manuscript).

[145] Kuhn, T. (1991a), The Road Since *Structure*', in A. Fine, M. Forbes, and L. Wessels (eds.), *PSA 1990*, Vol. II, Philosophy of Science Association, East Lansing, pp. 3-13.

[146] Kuhn, T. (1991b), The Natural and the Human Sciences', in D. R. Hiley, J. E. Bohman and R. Shusterman, eds. *The Interpretive Turn*, pp. 17-24, Ithaca: Cornell University Press.

[147] Kuhn, T. (1992), The Trouble with the Historical Philosophy of Science, Robert and Maurine Rothschild Distinguished Lecture 19 November 1991'. Cambridge Mass: An Occasional Publication of the Department of the History of Science, Harvard University.

[148] Kuhn, T. (1993), 'Afterwords', in P. Horwich (ed.), *World Changes*, MIT Press, Cambridge, pp. 311-341.

[149] Kunze, P.(1933), Z. Phys. 83, 1 (1933)

[150] Lakatos, I. and Musgrave, A. (eds.) 1970, *Criticism and the Growth of Knowledge*, Cambridge: Cambridge University Press

[151] Lakatos, I. (1970), "Falsification and the Methodology of Scientific Research Programmes", in Lakatos, I. and Musgrave, A. (eds.) 1970, *Criticism and the Growth of Knowledge*, Cambridge: Cambridge University Press, pp. 91-196.

[152] Lakatos, I. (1971a), "History of Science and Its Rational Reconstructions", in Buck, R. C. and Cohen, R. S. (eds.) 1971, PSA 1970 In Memory of Rudolf Carnap, Dordrecht: D. Reidel, pp. 91-136.

[153] Lakatos, I. (1971b), "Replies to Critics", in Buck, R. C. and Cohen, R. S. (eds.) 1971: 174-182.

[154] Lakatos, I., Mathematics, Science and Epistemology, Philosophical Papers, Vol. 2 (London/New York, 1978), ed. Worrall, J. et al.

[155] Lan Ju, Z. (1986a), "Resolving Power of Scientific Theory: A Brief Introduction to Pandit's View on the Appraisal of Scientific Theory", in Journal of Dialectics of Nature 2 (N0.4), pp. 1-10.(In Chinese).

[156] Lan Ju, Z. (1986b), "Epistemic Structuralism" - A translation from Pandit (1983) The Structure and Growth of Scientific Knowledge, in Philosophical Problems in Natural Science 3 (N0.3), pp. 4-10. (In Chinese).

[157] Lan Ju, Z. (1988), "A Systematic Introduction to Pandit's Philosophy", in Journal of Dialectics of Nature 10 (N0.4), pp. 9-16. (In Chinese).

[158] Lapid, Y. (1989), (Carleton University), 1989: "The Third Debate: On the Prospects of International Theory in a Post-Positivist Era", in International Studies Quarterly (1989) 33, 235-254.

[159] Lattes, C. M. G., Muirhead, H., G.P.S. Occhialini, and Powell, C. F. (1947), *Nature* 159 (1947) 694

[160] Laudan, L. (1977), *Progress and Its Problems*, Routledge and Kegan Paul: London.

[161] Laudan, L. (1987) "Progress or Rationality?: The Prospects for Normative Naturalism", American Philosophical Quarterly 24: 19-31.

[162] Lincoln, D. (2004), *Understanding the Universe: From Quarks to the Cosmos*, World Scientific: New Jersey/London/Singapore.

[163] London (1927) *"Quantenmechanische Deutung der Theorie von Weyl"* Z. Phys. 42.

[164] Mach, E. (1883), Die Mechanik in ihrer Entwicklung; historisch-kritisch dargestellt.

[165] Mach, E. (1896), Die Prinzipien der Wärmelehre, historisch-kritisch entwickelt.

[166] Mach, E. (1905), Erkenntnis und Irrtum : Skizzen zur Psychologie der Forschung. 1st German ed.

[167] Mach, E. (1976), Knowledge and Error: Sketches on Psychology of Enquiry. Vienna Circle Collection Volume 3. Dordrecht: D. Reidel. Translated by McCormack, T. J. et al.

[168] Mach, E. (1921), Die Prinzipien der physikalischen Optik historisch und erkenntnis-psychologisch entwickelt.

[169] Mckinney, W. J. (1993), "Review of G. L. Pandit, Methodological Variance: Essays in Epistemological Ontology and the Methodology of Science". xxiv + 422pp; figs., bibl., indexes. Kluwer Academic Publishers: Dordrecht/Boston/London, 1991. Isis 84, 1993, pp. 425-426.

[170] Mattingly, James (2005), "The Structure of Scientific Theory Change: Models versus Privileged Formulations", in *Philosophy of Science* 72, pp. 365-389.

[171] Maxwell, J. C. (1890), *Scientific Papers* Vol. 2, No. 61, Cambridge University Press.

[172] Maxwell, N.(1984), *From Knowledge to Wisdom*, Oxford: Basil Blackwell, 1984;

[173] Maxwell, N.(1989), *The Comprehensibility of the Universe: A New Conception of Science*, Oxford: Oxford University Press

[174] Maxwell, N. (2004), *Is Science Neurotic*, London: Imperial College Press, 2004.

[175] Maxwell, N. (2005) "Popper, Kuhn, Lakatos and Aim-Oriented Empiricism", *Philosophia* 32: 218-226, Nos. 1-4.

[176] Maxwell, N. (2006: 1-26), "Popper's Paradoxical Pursuit of Natural Philosophy", *Cambridge Companion to Karl Popper* (forthcoming)

[177] Maxwell, N. (2009), *What is Wrong with Science?*, second Edition, Pentire Press: London (first published by Bran's Head Books 1976).

[178] 169) Maxwell, N. (2010). "Reply to Comments on Science and the Pursuit of Wisdom", Philosophia (2010) 38;667-690.

[179] 170) Maxwell, N. (2012a), "Arguing for Wisdom in the University: An Intellectual Autobiography", Philosophia (2012) 40: 663-704.

[180] 171) Maxwell, N. (2012b), "In Praise of Natural Philosophy: A Revolution for Thought and Life", Philosophia (2012) 40: 705-715.

[181] Mielke, E. W. and Hehl, F. W. (1985) *Die Entwicklung der Eichtheorien, in Exact Sciences and their Philosophical Foundations*, ed. W. Deppert et al. (Hermann-Weyl-Kongress, Kiel).

[182] Miller, D. (1997) "Sir Karl Raimund Popper, C.H., F.B.A.- 28 July 1902 - 17 September 1994", *Biographical Memoirs of Fellows of the Royal Society* 43, 1997, 367-409.

[183] Moulines, C. U. (1995): "A Structuralist's Comment on Ludwig", in Lorenz Krueger and Brigitte Falkenburg (Hrsg.) *Physik, Philosophie und die Einheit der Wissenschaften: Fuer Erhard Scheibe* - Grundlagen der Exakten Naturwissenschaften Band 10. Spektrum Akademischer Verlag: Heidelberg/Berlin/Oxford, 1995, 300-305.

[184] Neddermayer, S. H., and C.D. Anderson, C. D., *Phys. Rev.* 51 (1937) 884

[185] Nersessian, Nancy J.: 2001, "Concept Formation and Commensurability" in Hoyningen-Huene, P., et al. (eds.): 2001, pp. 275-301.

[186] Newton, I. (1704): Opticks, Dover edition 1952, an unabridged and an unaltered republication of the 1931 edition of G. Bell and Sons Ltd., New York.

[187] Omn'es, R., 2002, Phys. Rev. A 65, 052119.

[188] Oppenheimer, J. R. (1937), and Serber, R., Phys. rev. 51 (1937) 1113

[189] O'Raifeartaigh, L. and Straumann, N. (2000): "Early History of Gauge Theories, with a Glance at Recent Developments", *Rev. Mod. Phys*, 72, 1-23.

[190] O'Raifeartaigh, L. (1997) The Dawning of Gauge Theories, Princeton University Press: Princetion, New Jersey.

[191] Osbeck, Lisa M. 2005: "Method and Theoretical Psychology" in Theory & Psychology, Vol. 15, No.1, pp. 5-26.

[192] Pais, A. (1982), Subtle is the Lord: *The Science and Life of Albert Einstein*. Oxford University Press: New York.

[193] Pais, A. (1986), *Inward Bound*. Oxford.

[194] Pandit, G. L. (1971a): *Physical Theory: Explicating Structure and Meaning*. Ph. D thesis written during 1966-1971 in the analytic tradition of the twentieth century philosophy of science (unpublished) focusing on *Problems of Structure and Content* with reference to the work of the most leading analytic philosophers of twentieth century: Rudolf Carnap, Moritz Schlick, Carl Hempel, Karl Popper and Paul Feyerabend.

[195] Pandit, G. L. (1971b), "Two Concepts of Psychologism", *Philosophical Studies*, International Journal for Philosophy in the Analytic Tradition 22, 85-91.

[196] Pandit, G. L. (1982), *The Structure and Growth of Scientific Knowledge: A Study in the Methodology of Epistemic Appraisal.* Boston Studies in the Philosophy of Science Vol. 73, D. Reidel: Dordrecht/Boston/London.

[197] G. L. Pandit (1983), "Psychology or Logic of Inquiry", in Indian Philosophical Quarterly 10 (1983, 393-401). A Review Essay based upon Ernst Mach: 1976, Knowledge and Error: Sketches on Psychology of Enquiry, Dordrecht: D. Reidel. Vienna Circle Collection 3, translated from the original German edition (Mach 1905) by T. J. McCormack and Paul Foulkes

[198] Pandit, G. L. (1986), "Rationality of an Optimum Aim for Science", Review of J. Watkins (1984) *Science and Scepticism*, Hutchinson: London, in *Journal of Indian Council of Philosophical Research* **3**(Spring 1986, 141-148).

[199] G. L. Pandit (1987), "Epistemological Ontology and the Special Sciences: An Interaction-Theoretic Argument Against Relativism",in The Journal of the Indian Council of Philosophical Research 4 (35-45).

[200] Pandit, G. L. (1988), "Science and Truthlikeness", The Journal of the Indian Council of Philosophical Research 5 (125-138).

[201] Pandit, G. L. (1989), "Scientific Change: The Possibility of a Unified Approach", in Paul Weingartner und Gerhard Schurz (Herausgeber) *Grenzfragen zwischen Philosophie und Naturwissenschaften* (Schriftenreihe der Wittgenstein-Gesellschaft Band 18), Wien: Hoelder-Pichler-Tempsky, 1989, S. 168-179.

[202] Pandit, G. L. (1990), "Scientific Change: The Search for the Methodology of Rational Theory-Choice", in World Affairs 1 (112-118).

Pandit, G. L. (1991), *Methodological Variance: Essays in Epis- temological Ontology and the Methodology of Science.* Boston Studies in the Philosophy of Science Vol. 131, Springer/Kluwer Academic Publishers: Dordrecht/Boston/London.

[203] Pandit, G. L. (1993), "Die andere Seite des Wissenschaftlichen Realismus", in Neue Realitaeten: Herausforderung der Philosophie - XVL. Deutscher Kongress fuer Philosphie, T. U. Berlin, 20.-24.September 1993, S. 810-817.

[204] Pandit, G. L. (1994a), Die Primordialitaet und die Wirklichkeit der Sprache bei Wittgenstein (unpublished).

[205] Pandit, G. L. (1994b), "Paul Feyerabend (1924-1994)", in Journal of the Indian Philosophical Research 11 (1994, 167-168).

[206] Pandit, G. L. (1994c), "Karl Raimund Popper (28. July 1902 - 17. September 1994)",in The Journal of the Indian Council of Philosophical Research 12 (1994, 189-192).

[207] Pandit, G. L. (1995): "A Pre-Established Dis-Harmony?" (1995: 152-162), A Response to Nancy Cartwright, "Where in the World is the Quantum Measurement Problem?", in Lorenz Krueger and Brigitte Falkenburg (Hrsg.) *Physik,*

Philosophie und die Einheit der Wissenschaften: Fuer Erhard Scheibe - Grundlagen der Exakten Naturwissenschaften Band 10. Spektrum Akademischer Verlag: Heidelberg/Berlin/Oxford, 1995, 130-151.

[208] Pandit, G. L. (1996), "Lorenz Krueger: Rational Reconstructionist of Conditions of Scientific Change", in The Memorial Symposium for Lorenz Krueger Max-Planck-Institut fuer Wissenschaftsgeschichte: Berlin: 25th September 1995. Preprint 38 (1996, 27-35).

[209] Pandit, G. L.(1998), "Review of Samuel P. Huntington, *The Clash of Civilizations and the Remaking of World Order*, Viking: Penguin India, 1997, pp. 367, in *World Affairs - The Journal of International Issues*, Vol. 2 (Spring 1998), pp. 135-137.

[210] Pandit, G. L. (1999), "L'impact de la pensee de Sir Karl Popper sur notre comprehension du monde naturel et du monde des homme", Quo vadis: Revue Internationale de Philosophie 35 (1999, 82-94).

[211] Pandit, G. L. (2002a), "Heisenberg-Einstein Context Principle and the Dynamic Core-Context of Discovery in Physics", in Fortschritte der Physik 50: 5-7, 461-482.

[212] Pandit, G. L. (2002b), "Heisenberg-Einstein Context Principle and the Dynamic Core-Context of Discovery in Physics", in D. Papenfu, D. Luest, W. P. Schleich (eds.) *100 Years Werner Heisenberg - Works and Impact*. Wiley-VCH, pp. 32-53.

[213] Pandit, G. L. 2003. "The Work and the Pilgrims of Music", in Andreea Deciu Ritivoi (ed.) Interpretation and Ontology: Studies in the Philosophy of Michael Krausz, 2003, 293-302. Rodopi: Netherlands (Value Inquiry Series Vol. 146).

[214] Pandit, G. L. (2007a), "Epistemologically Embedded Methodology of Science: Turns in the Twentieth Century Conceptions of Scientific Rationality", in B. V. Sreekantan (ed.) Foundations of Sciences. PHISPC Publications Vol. XIII, Part 5, pp. 43-73, New Delhi (forthcoming)

[215] Pandit, G. L. (2007b), Quantum Core-Context Rebuilding: The Methodology of Theory-Problem Interactive Systems in Physics (in preparation)

[216] Pandit, G. L. (2007c) "Rethinking Science-Studies with a Return to Natural Philosophy", *Friends of Wisdom News Letter* (No. 1, Nov. 2007: 28-32), reviewing Nicholas Maxwell: *Is Science Neurotic?* Imperial College Press: 2004, pp.xx+240. For details, visit FOW archive at: www.jiscmail.ac.uk/lists/friendsofwisdom.html.

[217] Pandit, G. L. (2008a), "Universities with a Room for Wisdom Inquiry", in Friends of Wisdom Newsletter, No. 3, July 2008, pp. 4-9.

[218] Pandit, G. L. (2008b) "Where in the World of Science are there Paradigms-in-Charge or Paradigms-in-Crises?", presented as an invited contribution at the International Symposium on "Incommensurability Thesis" held at the Central University of Hyderabad, India (18-20 January 2007), unpublished.

[219] Pandit, G. L. (2010a), "Aim-Oriented Empiricism: How We Might Improve the Aims and Methods of Science, Making it More Rational, Responsible and Far-Reaching", in Friends of Wisdom Newsletter No. 6, january 2010, pp. 30-37.

[220] Pandit, G. L.(2010b), "How Simple is it for Science to Acquire Wisdom According to its Choicest Aims", Philosophia 38 (2010; 649-666).

[221] Passmore, John (1985), *Recent Philosophers: A Supplement to A Hundred Years of Philosophy*, Duckworth: London.

[222] Pauli, W. (1921), "Relativitätstheorie" *Enz. Der math. Wiss.*, Dd. 5.

[223] Pauli, W. (1933), *Die allgemeinen Prinzipien der Wellenmechanik*. Handbook der Physik, Berlin

[224] Pierce, A. (2008), "Dark Matter at the LHC", in Kane, G. & Pierce, A. (2008), pp. 13-23

[225] Pickering, A. (1984) Constructing Quarks: A Sociological History of Particle Physics, Chicago: University of Chicago Press.

[226] Planck, M. (1965), Vorträge und Erinnerungen, Wissenschaftliche Buchgesellschaft: Darmstadt (reprinted from 5th ed. Stuttgart 1949).

[227] Poincare, H. (1902), Science et l'hypothese, Paris 1902

[228] Poincare, H. (1905), Le valeur de la science 1905.

[229] Poincare, H. (1909), Science et Methode, Paris 1909.

[230] Popper, K. R.: (1934), *Logik der Forschung*, Wien 1934 (expanded English edition: 1959).

[231] Popper, K. R. (1945) The Open Society and its Enemies. Routledge & Kegan Paul.

[232] Popper, K. R. (1957) The Poverty of historicism, Routledge & Kegan Paul

[233] Popper, K. R. (1959), *The Logic of Scientific Discovery* (trans. Logic der Forschung, Wien 1934) London: Hutchinson/Basic Books Inc: New York

[234] Popper, K. R.: (1963), *Conjectures and Refutations*, London: Routledge and Kegan Paul, 1963.

[235] Popper, K. R.: (1970), "Normal Science and Its Dangers" in Lakatos, I. et al (1970: 51-58).

[236] Popper, K. R. (1972a), *Conjectures and Refutations*, London: Routledge and Kegan Paul (first ed. 1963).

[237] Popper, K. R. (1972b), *Objective Knowledge: An Evolutionary Approach*, London: Oxford University Press.

[238] Popper, K. R. (1982), The Open Universe: An Argument for Indeterminism. From the Postscript to Logic of Scientific Discovery. Hutchinson. Ed. by W.W. W. Bartley, III.

[239] Popper, K. R. (1983), *Realism and the Aim of Science*, from the Postscript to the Logic of Scientific Discovery, Vol. III, ed. by W. W. Bartley, III. London/New York: Routledge. Paperback ed. 1985

[240] Popper, K. R. (1994), Alles Leben ist Problemloesen. Piper: Muenchen/Zuerich.

[241] Popper, K. R.(1974), Unended Quest: An Intellectual Autobiography (Fontana, 1976), first published as "Autobiography of Karl Popper" in: The Philosophy of Karl Popper, ed. P. A. Schilpp (Illinois, 1974).

[242] Quine, W. V. (1960), *Word and Object*, Cambridge: MIT Press.

[243] Quine, W. V. (1974), *Roots of Reference*, LaSalle, IL: Open Court Publishers.

[244] Quine, W. V. (1981), *Theories and Things*, Cambridge: MIT Press

[245] Quine, W. V. (1995a), "Two Dogmas in Retrospect", *Canadian Journal of Philosophy* 21, 265-274.

[246] Quine, W. V. (1995b), *From Stimulus to Science*, Cambridge: Harvard University Press.

[247] Raby, S. (2004) "Grand Unified Theories", *Physics Letters B: Review of Particle Physics* 592: 186-190. Issues 1-4

[248] Randall, Lisa, (2005), *Warped Passages: Unravelling the Mysteries of the Universe's Hidden Dimensions*. Ecco.

[249] Salam, A. (1968), *Elementary Particle Theory*, ed. N. Svartholm (Almquist and Wiksell, Stockholm, 1968).

[250] Scheibe, E. (2001), *Between Rationalism and Empiricism: Selected Papers in the Philosophy of Physics*. Springer Verlag: New York. PP. 136-141.

[251] Scheibe, E., "The Physicists Conception of Progress", Studies in History and Philosophy of Science 19 (1988), pp. 141-159.

[252] Scherk, J. and Schwarz, J. H. (1974) "Dual Models for Non-Hadrons", Nucl. Phys. B 81 (1974) 118-144.

[253] Schilpp, Paul Arthur. 1963/1987 (ed.) The Philosophy of Rudolf Carnap. The Library of Living Philosophers, INC./La Salle : Illinois. Open Court. London : Cambridge University Press.

[254] Schlick, M. (1949), "Meaning and Verification", in Feigl, H. et al., eds. Readings in Philoso-phical Analysis. New York: 1949. Rep. from Schlick, M. "Meaning and Verification", The Philosophical Review 45, 1936.

[255] Schmitz-Rigal, Christiane (2002), Die Kunst offenen Wissens, Hamburg.

[256] Schmitz-Rigal, Christiane (2003), Die Kunst der Wissenschaft, in Philosophia Naturalis Band 40/ Heft 2, S.255-291.

[257] Schroedinger, E. (1922) "Über eine bemerkenswerte Eigenschaft der Quantenbahnen eines einzelnen Elektrons" Z. Phys. 12.

[258] Schweber, S. (1949) QCD and the Men who made it, Princeton.

[259] Seelig, C. (1960), Albert Einstein. Zuerich: Europa Verlag: Zürich.

[260] Siegel, H. (2001):"Incommensurability, Rationality and Relativism: In Science, Culture and Science Education" in Hoyningen-Huene, P., et al. (eds.): 2001, pp. 207-224.

[261] Seiler, E. et al. (eds.) 2007, Approaches to Fundamental Physics: An Assessment of Current Theoretical Ideas, Springer Verlag: Berlin/Heidelberg/New York

[262] Seiler, E. et al.: 2007, Introduction - The Many-Fold Way of Contemporary High Energy Theoretical Physics in Erhard Seiler et al. (eds.) Approaches to Fundamental Physics, Lecture Notes in Physics 721, Springer, pp.3-18.

[263] Sieroka, N.(2010), *Umgebungen*, Chronos, Zürich 2010

[264] Söding, P. Eur. Phys. Journal H **25** (2010) 3

[265] Stachel, J. (1994): "Scientific Discoveries as Historical Artifacts" in Kostas Gavroglu et al. (eds.), Trends in the Historiography of Science, Boston Studies in the Philosophy of Science 151, Kluwer Academic Publishers: Dordrecht/Boston/London, pp. 139-148.

[266] Stegmüller, W. (1979), The Structuralist View of Theories, Springer Verlag: Berlin/Heidelberg/New York

[267] Straumann, N. (1987) "Zum Ursprung der Eichtheorien bei Hermann Weyl" Phys. Bl. 43, Nr. 11, pp. 414-421

[268] Suppe, Frederick (1971), "On Partial Interpretation", in The Journal of Philosophy Vol. LXVIII, No.3, Feb. 11, pp. 57-76.

[269] Suppe, Frederick (1972), "Theories, Their Formulations, and the Operational Imperative", in Synthese 25, Nos.1/2, pp. 129-164.

[270] Suppe, Frederick (1977), The Structure of Scientific Theories (1st ed. 1974). Chicago: University of Illinois Press

[271] Suppe, Frederick (1989), The Semantic Conception of Theories and Scientific Realism, Chicago: University Of Illinois Press.

[272] Swanson, J. W. (1967), Discussion on the D-thesis', Philosophy of Science 34, pp. 59-68

[273] Taylor, Jr., J. H. (1993), "Binary Pulsars and Relativistic Gravity", Nobel Lecture, Dec. 8, Physics 1993 (73-91).

[274] Toulmin, S., "From Logical Systems to Conceptual Populations" in: PSA 1970: In Memory of Rudolf Carnap, ed. R. C. Buck et al. (Dordrecht, 1971), pp. 552-564

[275] Truesdell, C. (1980), The Tragicomical History of Thermodynamics 1822-1845 New York Heidelberg/Berlin.

[276] Valtonen, M. J. et al. (2008), "A Massive Binary Black-Hole System in OJ287 and A Test of General Relativity", Nature 452, 17th April (2008: 851-853)

[277] van Frassen, Bas C. (1980) The Scientific Image. Oxford: Clarendon Press;

[278] van Fraassen, Bas C.(1987), "The Semantic Approach to Scientific Theories", in Nancy J. Nersessian (ed.), Science and Philosophy. Dordrecht: Martinus Nijhoff Publishers.

[279] Van Frassen, Bas C. (1989) Laws and Symmetry, Oxford University Press.

[280] Van Fraassen, Bas C. (1991) Quantum Mechanics: An Empiricist View. Oxford: Clarendon Press

[281] Veltman, Martinus (2003), Facts and Mysteries in Elementary Particle Physics, Paperback, World Scientific Publishing Company.

[282] Veneziano, G. (1968) "Constructon of a Crossing Symmetric", Regge-Behaved Amplitude for Linearly Ris Nuov. Cim., 57A: 190.

[283] Wartofsky, M. W., "The Relation Between Philosophy of Science and History of Science" in: Essays in Memory of Imre Lakatos, ed. Cohen, R. S. et al. (Dordrecht, 1976), pp. 717-737.

[284] Watkins, J. (1984) Science and Scepticism, Hutchinson: London.

[285] Wegner, F. (1971), J. Math. Phys.,12, 2259 (1971)

[286] Weinberg, S. (1967) "A Model of Leptons", Phys. Rev. Lett. 19 (1967), 1264

[287] Weinberg, S. (1992) Dreams of a Final Theory. Pantheon Books: New York.

[288] Weinberg, S. (1997) "Changing Attitudes and the Standard Model" (36-44), in Lillian Hoddeson, Laurie Brown, Michael Riordin and Max Dresden (eds.) The Rise of the Standard Model: Particle Physics in the 1960's and 1970's. Cambridge University Press.

[289] Weyl, H. (1918a), Raum, Zeit, Materie. Springer: Berlin;

[290] Weyl, H. (1918b) "Reine Infinitesimalgeometrie" Math. Z., 2, 384-411.

[291] Weyl, H. (1919) "Eine neue Erweiterung der Relativitätstheorie", Ann. Phys. 59 (101-133).

[292] Weyl, H. (1922) Raum, Zeit , Materie. Berlin.

[293] Weyl, H. (1929) "Elektron und Gravitation" SZ. f. Phys. 56 (330-352). Reprinted in H. Weyl (1968) Gesammelte Abhandlungen III. p. 245.

[294] Weyl, H. (1949) "Wissenschaft als symbolische Konstruktion des Menschen", Eranos-Jahrbuch 1949, p.375. Reprinted in Weyl (1968)

[295] Weyl, H. (1952), *Symmetry*, Princeton University Press, Princeton, 1952.

[296] Weyl, H. (1968) Gesammelte Abhandlungen, Springer-Verlag: Berlin/Heidelberg (Ed. K. Chandrasekharan). Berlin/Heidelberg/New York 1968

[297] Wheeler, J. A., "Niels Bohr: The Man and His Legacy" in: The Lesson of Quantum Theory, ed. de Boer, J. et al. (Amsterdam, 1986), pp. 355-367.

[298] Wigner, E. P. (1960) "The Unreasonable Effectiveness of Mathematics," Communications in Pure and Applied Mathematics 13 (1960): 1-14.

[299] Wigner, E. P. (1971) "The Subject of Our Discussions", B. d'Espagnat (ed.) Foundations of Quantum Mechanics. International School of Physics "Enrico Fermi" 1970, Academic Press, New York, 1971, pp. 1-9.

[300] Wigner, E. P. (1982), "The Limitations of the Validity of Present-Day Physics" (pp. 118-133), in Richard Q. Elvee (ed.) Mind in Nature: Nobel Conference XVII. Harper & Row Publishers, San Francisco, 1982, p. 124.

[301] Wilson, K.G. (1974), Phys. Rev., D10, 2445 (1974)

[302] Wittgenstein, L. 1922. Tractatus Logico-Philosophicus. English translation with an Introduc- tion by G. E. M. Anscombe: Wittgenstein, L. 1959.Tractatus Logico-Philosophicus. London: Hutchinson University Library

[303] Wittgenstein, L. (1953), Philosophische Untersuchungen (PU) 1953.

[304] Wittgenstein, L. (1953), Philosophical Investigations. Basil Blackwell: Oxford.

[305] Wolszczan, A. (1991), "A nearby 37.9 ms radio pulsar in a relativistic binary system" Nature, 350: 688-690, 1991.

[306] Yang, C. N. (1986) "Hermann Weyl's Contribution to Physics", in Hermann Weyl 1885-1955: Centenary Lectures, Springer-Verlag: Berlin/Heidelberg/NewYork, pp. 7-21. Ed.by K. Chandrasekharan.

[307] Yang, C. N. (1983), in Proc. Int. Sym. Foundations of Quantum Mechanics: Tokyo. Ed.by S. Kamefuchi, H. Ezawa et al. P. 5 (Phys. Soc. of Japan, 1984).

[308] Yang, C. N. and Mills, R.(1954) Phys. Rev. 96.

[309] Yang, C. N. and Mills, R.(1954) Phys. Rev. 95.

[310] Yoneya, T. (1974) "Connection of Dual Models to Electrodynamics and Gravidynamics", Prog. Theor. Phys. 51 (1974) 1907-1920.

[311] Yukawa, H. (1935) Proc. Phys. Math. Soc. Japan.17 (1935), 48

[312] Zurek, W. H., (2003), Rev. Mod. Phys. 75, 715

[313] Zwicky, F., (1933) "Die Rotverschiebung von extragalaktischen Nebeln", Helvetica Physica Acta 6, 110-127

Abstract

This book addresses the question of how to articulate the scientific framework of problem and theory development, taken in its dynamic complexity, at the active frontiers of physics. This poses a challenge because, in our view, the traditional explications of scientific progress and growth of knowledge simply in terms of the logical or meta-theoretical framework of theory-justification or theory-refutation do not address this question. The same is true of explications sought in terms of the aims of scientific inquiry, themselves chosen from outside science. The book shows how dynamic core-context building from within physics itself shapes the structure of scientific reasoning that nourishes the inner development of physics. Thus, it seeks to articulate theory development and theory-unification by arguing how within physics itself scientific reasoning from the dynamic core-context building resolving power of physical theory plays a methodological role in determining and shaping the frontiers of problem and theory development.

www.ingramcontent.com/pod-product-compliance
Lightning Source LLC
Chambersburg PA
CBHW081431170526
45166CB00008B/2173